SYSTEM-ON-CHIP METHODOLOGIES & DESIGN LANGUAGES

T0205365

SYSTEM-ON-CHIP METHODOLOGIES & DESIGN LANGUAGES

Edited by

Peter J. Ashenden

Ashenden Designs Pty. Ltd., Australia

Jean P. Mermet

ECSI, France

and

Ralf Seepold

Universität Karlsruhe, Germany

KLUWER ACADEMIC PUBLISHERS
BOSTON / DORDRECHT / LONDON

A C.I.P. Catalogue record for this book is available from the Library of Congress.

Published by Kluwer Academic Publishers,
P.O. Box 17, 3300 AA Dordrecht, The Netherlands.

Sold and distributed in North, Central and South America
by Kluwer Academic Publishers,
101 Philip Drive, Norwell, MA 02061, U.S.A.

In all other countries, sold and distributed
by Kluwer Academic Publishers,
P.O. Box 322, 3300 AH Dordrecht, The Netherlands.

Printed on acid-free paper

ISBN 978-1-4419-4901-1

Printed in the Netherlands.

TABLE OF CONTENTS

Contributors

P. Alexander, W. Anderson, T. Anderson, M. Andrews, M. Antoniotti, P. Ashenden, J. Aylor, B. Bailey, D. Barton, T. Bartos, D. Borrione, C. Brandolese, O. Bringmann, E. Cerny, G. de Jong, W. Ecker, L. Entrena, P. Faye, A. Ferrari, A. Flesca, N. Fristacky, P. Georgelin, M. Glesner, F. Hamid, C. Hansen, X. Hongxi, A. Jantsch, B. Jinian, B. Jinsong, M. Josephs, J-Y. Jou, G. Jun, L. Kabous, J. Kacerik, T. Kazmierski, R. Kazumiti Morizawa, R. Klenke, S. Krolikoski, D. Landoll, C. Lennard, H-M. Lin, C. López, J. Mades, G. Martin, N. Martínez Madrid, P.Menchini, D. Monjau, V. Moraes Rodrigues, T. Nanya, W. Nebel, F. Oppenheimer, G. Peterson, A. Prihozhy, W. Putzke-Röming, M. Radetzki, T. Riesgo, W. Rosenstiel, B. Salefski, I. Sander, A. Sangiovanni-Vincentelli, P. Schaumont, F. Schirrmeister, T. Schneider, G. Schumacher, R. Seepold, S. Seshadri, J. Stinson, S. Switzer, S. Thiyagarajan, Y. Torroja, J. Uceda, S. Voges, R. Waxman, J. Willis, A. Windisch, W. Wu, H. Xianlong, F. Yiping,

Preface

The Forum on Design Languages (FDL) is *the* European forum for the exchange of ideas and the place to learn about new directions and trends in design languages and methodologies – with a focus on *design reuse* and *system level design*. FDL has become established as a multi-faceted event, offering a number of related topics that make it efficient and easy to get (and give) up to date opinions, overviews and information.

FDL2000 is the third in this series of conferences and, after Lausanne and Lyon, Tübingen has been chosen to host FDL2000 in Germany. The forum book summarizes three main events: the HDL Workshop / VHDL Users' Forum in Europe (HDL&VUFE), the Workshop on Virtual Component Design & Reuse (VCDR) and the Workshop on System Specification & Design Languages (SSDL). Traditionally, HDL&VUFE gathers together VHDL and other HDL users and provides a complete snapshot of the status of practical usage of VHDL and other hardware description languages in the electronic design community. VUFE is the European VHDL meeting on HDL related topics and standardization efforts. The VCDR part is the successor of the International Workshop on Reuse Techniques for VLSI Design that was founded in 1996. This annual event is dedicated to a broad spectrum of digital and analog reuse of virtual components. The main objective is to present new ideas and methodologies for reuse and IP. Furthermore, standardization is regarded as a crucial issue. SSDL, which was started in 1996, addresses the need to develop a general consensus on key problems met by the designers of System on a Chip applications. These problems relate - among others - to the description of specifications, functional and implementation constraints and the usability of EDA tools.

The selection of high-quality contributions may help to represent a detailed overview on relevant items and to cover the main domains from the three areas mentioned. Thus supporting to exploit recent results in educational and industrial environments.

Ralf Seepold
FDL 2000 General Chair

The Annual International Hardware Description Language Conference and Exhibition (HDLCON) is the premier HDL event in the United States, held in Spring in the Silicon Valley area of California. It originated in the International Verilog Conference sponsored by Open Verilog International (OVI) and the Spring VHDL International User's Forum (VIUF) Conference sponsored by

VHDL International. Both of these events addressed not only language-specific issues, but also general issues of design methodology and design experience. Given the increasing common interest, the two conferences co-located in 1997 and 1998, and in 1999 they merged to become the HDLCON that we enjoy today. The scope of the conference has also expanded from Verilog and VHDL to include emerging HDL such as SystemC and system-level design languages such as Rosetta.

This volume includes a selection of the papers from HDLCON-2000 identified by the Program Committee as being high-quality papers in their areas. The topic area are system level design, mixed-language design, formal verification, design reuse, language standards, test-benches and verification, hardware/software co-design, and design tools. These are all important topics for system-on-a-chip designers, and the inclusion of the papers here will help disseminate the research and experiences to the design community.

Peter Ashenden
HDLCON-2000 Program Chair

APCHDL was created in 1993 in Brisbane Australia. I had the honor to be the General co-chair of this very successful event. I remember to have been requested to make a small speech at the dinner party that took place during a boat river cruise. While dreaming to the future, I then launched the idea of an annual event that would be hosted successively by the major Asian nations. My dream became reality along the following years when the conference happened successively in Japan, in India, in Taiwan, in Korea and finally in 2000 in mainland China.

This conference is an opportunity to the excellent young Asian researchers who are lacking, most of the time the means to travel to foreign events, to meet with the scientists of a domain – Chip Design Languages- which was not traditionally well developed in Asia. It is a great satisfaction to see that the quality of contributions is fully comparable to the quality of FDL or HDLCon papers. I have therefore the pleasure to introduce in this book a selection of the best papers presented in Beijing during August 2000. These presentations will complement nicely the equivalent selections coming from the 2 sister European and American conferences.

In conclusion, I must recall that the idea of the CHDL series comes from the IFIP10.5 Working group.

Since 1975 the Conference on Hardware Description Languages and their applications (CHDL) had been the flagship of this group. Then, it appeared in 1999 that a worldwide conference traveling every other year from a continent to the other would be less efficient than 3 regional events occurring every year in the 3 main regions. It was then suggested to keep the memory of the CHDL through a new series of books that would put together the best papers of all 3 events in the domain of Chip Design Languages (ChDL). I hope that a large public will take advantage of this worldwide harvest.

Jean Mermet
APChDL 2000 Program Chair
IFIP WG 10.5 Chair

VHDL TRENDS

VHDL in 2005 - The Requirements

Robert H. Klenke, James H. Aylor, Paul Menchini, Ron Waxman, William Anderson, Jack Stinson
Virginia Commonwealth University, University of Virginia, Menchini & Associates, EDA Standards Consulting, Advanced Technology Institute, Advanced Technology Institute

Key words: VHDL, System-level, IP, EDA tools

Abstract: The ability to put tens or hundreds of million transistors on a chip is no longer limited to just the developers of large microprocessors. Even ASIC designers at small companies have access to this technology and can develop "systems-on-a-chip" for their specific application. This capability is pushing current design methodologies and tools beyond the limit. Advancements in design technology must be developed and exploited to keep designer productivity in line with the advancements in fabrication technology. A key element in the development of new design technologies is the use of hardware description languages. VHDL was originally designed to describe systems from an abstract level to the gate level, but its use in current design methodologies is straining to keep up with the size and complexity of systems being developed. A continuous improvement process is necessary to ensure that VHDL has the features and capabilities required to implement the design methodologies of the future. This paper describes the results of a user survey of designers who use VHDL. This survey is intended to be a first step in identifying the requirements for the next update to VHDL - VHDL 200X. The next update to VHDL must provide the enhancements to the language necessary to take it into the next decade as the premier language for the design of systems-on-a-chip.

1. INTRODUCTION

In 1998, the Electronics Directorate of the U.S. Air Force funded a project to assist the VHDL user community in developing the next update to

P.J. Ashenden et al. (eds.), System-on-Chip Methodologies & Design Languages, 3–12.
© 2001 *Kluwer Academic Publishers. Printed in the Netherlands.*

the VHDL language - VHDL 200X. The authors, as the project team, developed a plan to gather requirements for the VHDL 200X update. This plan, shown in Figure 1, was based, in part, on a survey of the VHDL user community at-large to determine their needs for VHDL enhancements. This survey was planed to consist of two phases. Phase I was a web-based survey which was advertised to as broad a membership of the user community as possible. The goal of Phase I was to gather general information from a large population of users to help compile a list of user needs. The needs identified will be used to develop a more focused, detailed survey of a smaller subset of users. The results of the Phase II survey will help the VHDL language development community (the IEEE CS VASG) to make a significant enhancement in the capabilities of VHDL for the year 200X, well beyond the "maintenance release" which is in process today.

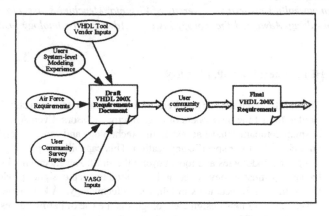

Figure 1. VHDL 200X requirements development plan

This paper presents the data from the Phase I user survey conducted by the VHDL 200X team.

2. USER COMMUNITY GUIDANCE FOR REQUIREMENTS DEVELOPMENT

The requirements depicted below are "soft" requirements, in the sense that they actually provide guidance for the development of "hard" or specific language requirements. Specific language requirements should be stated in a way that aids the language developer and tool builders to implement the recommended requirements.

The survey results were broken down into the major categories described below. In each section, discussion of the data received and some initial conclusions based on that data are included. These conclusions are presented *only* as a means to start the discussion of requirements for VHDL 200X - they are not intended to be seen as facts or hard-and-fast opinions by the team.

2.1 Demographics

This section of the survey included questions on the respondent's major job description, their companies' product area, and their use of HDLs. From the data, it appears that the majority of the respondents are VHDL users, but not Verilog users. This is somewhat expected from the distribution of the survey announcement. However, we were expecting more of a mix in that most tools today support both languages in cosimulation and the user community appears to be more bilingual, having to work with systems where components are modeled in VHDL and Verilog. Perhaps the low percentage of Verilog users can be explained by the focus of the survey - primary Verilog users did not feel compelled to answer questions on a VHDL survey. One very interesting result is the significant percentage of respondents who claimed to be using C or C++ as a hardware description language as shown in Figure 2. It is not clear if this use is in modeling hardware itself or in modeling the software that is a major part of most systems-on-a-chip being currently developed. Further investigation of this issue is planned for the second phase of the survey.

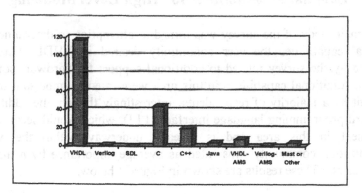

Figure 2. What HDL do you use?

Another fact shown in the survey data was that the majority of the respondents indicated that the most of their time was spent in detailed design, simulation, and verification from the RTL level down to gates and a final implementation as shown in Figure 3.

6

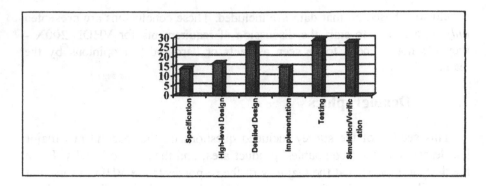

Figure 3. Where in the design cycle (specification, high-level design, detailed design, implementation, testing, etc.) is the majority of time spent in your organisation?

This result was further reinforced by the data from an essay question in the survey where a majority of respondents felt that the most important application area for VHDL currently is the detailed design of ASICs and FPGAs. Obviously this is currently the case, but we must insure that we gather requirements from the user community that is looking past this into the future where system design at higher levels of abstraction will become a very important application domain for VHDL.

2.2   Language Capabilities for High Level Modeling

The main body of the survey was aimed at attempting to determine what additional capabilities the user community desired in VHDL. One area addressed by the survey related to additional support for hardware/software codesign. Additional capabilities in this area were rated as important or very important by a majority of respondents. Interestingly though, the addition of a standard programming language interface (PLI), which would seem to be a requirement in this area and is already underway within the VHDL standardization community, was rated as average importance by a majority of respondents. These results are shown in Figure 4 below.

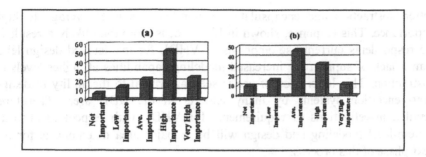

Figure 4. (a) How important is Hardware/Software codesign support? (b) a standard programming language interface by which externally generated (foreign, e.g. C, C++) models and functions can be included in the simulation of a design?

Further investigation of the user's needs in this area and why perhaps they do not feel that a standard PLI is a part of hardware/software codesign support will be one of the focus areas of the phase II survey.

Another area that was covered in the survey was that of "system-level" design, or designing at higher levels of abstraction. Within that area, the respondents indicated that the capability to express system-level constraints for power, timing, etc. was very important. On the other hand, they indicated that an interface to a high level language such as the System Level Design Language (SLDL), which currently is concentrating on the area of constraint specification was only of average importance. This apparent conflict, shown in the graphs of Figures 5, may be the result of the user's desire to express speed, area and power constraints for synthesis within the language itself instead of as separate inputs to the synthesis tools as is now required.

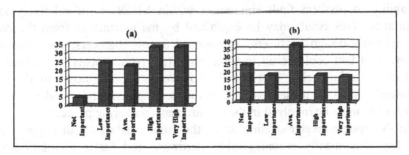

Figure 5. (a) How important is system-level constraint specification within VHDL (power, timing constraints, etc.)? (b) an ability to interface to high-level system description languages (SDL, SLDL)?

Also in the area of system-level modeling, the respondents indicated that system-level performance and stochastic modeling and better communications modeling capabilities were of average importance but that better abstraction and encapsulation capabilities were of average to high importance. This response, shown in Figure 6, is more than likely a result of the respondents current use emphasis of VHDL for the detailed design level than a lack of support for increased modeling capabilities at higher levels of abstraction. This latter area of use is seen as critical to the ability to design next-generation systems by many experts in the HDL area. Therefore, detailed investigations of requirements for VHDL with respect to abstract system-level modeling and design will be a major area of emphasis for the next phase of this process.

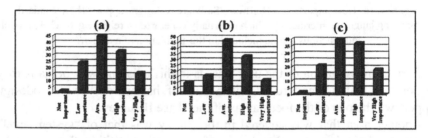

Figure 6. (a) How important is system level modelling (e.g. performance and stochastic modelling)? (b) are better communications modelling capabilities? (c) are better encapsulation and abstraction capabilities?

The respondents to the survey indicated that high-level synthesis from the behavioral level to the RTL level is important and that is generally the impression one gets from the tool vendors as well. Exactly what requirements this area may generate in the language, if any is unclear, but will be investigated. In addition, a majority of the respondents indicated that the ability to perform fault simulation within VHDL is of relatively low importance. This result may be explained by the indications from the data that primarily design engineers, as opposed to test engineers, answered the survey. This fact could result in some important requirements in the area of production testing being overlooked. To avoid this, engineers in the test community, such as those involved in the development and standardization of WAVES, must be involved in the requirements gathering process.

In the system-level modeling area, the survey asked about support for modeling dynamically reconfigurable hardware and faster simulation via more efficient communications mechanisms, both of which were rated as average importance. Perhaps the dynamically reconfigurable computing area is still too immature to generate much interest and the users may be unaware of the benefits that modeling communications without using VHDL signals would bring. However these might be very fruitful areas of use for exploitation of VHDL and should be looked at in more detail.

Finally, the survey results, shown in Figure 7, pointed out that a majority of the community seems to be unaware of the currently ongoing work in the areas of Object Oriented VHDL (OOVHDL) and System and Interface Design (SID) and how those techniques may help designers in the future. It is clear from this data that groups that are involved in developing new language extensions should ensure that the community is informed of their work. Furthermore, the requirements expressed by the users must be "filtered" in such a way that the requirements that will be addressed by OOVHDL and SID are not duplicated.

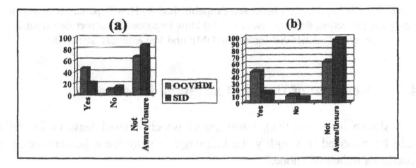

Figure 7. (a) Are you aware of the following extensions to VHDL currently being pursued and do you feel they will be useful in addressing some of the capabilities described [in the survey]? (b) Do you feel they will be useful to YOU in the future?

2.3 Language Capabilities for Intellectual Property (IP) Generation and Reuse

Several questions in the survey asked about language support for IP generation and reuse. The results are shown in Figure 8. The respondents indicated that capabilities for reuse were of high importance, but what specific requirements exist outside of the areas of encapsulation and abstraction that may be addressed by OOVHDL are unclear. Support for IP creation, protection, delivery and use was rated as high to very important. However, specific requirements in this area are unclear, especially since the support of emerging standards in this area such as VSI and OMF were rated as average to low importance. Clearly, this is an area where more study is needed.

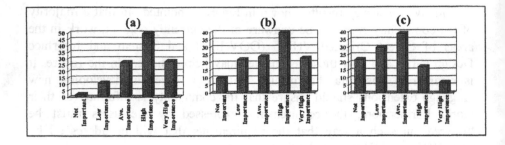

Figure 8. (a) How important are better reuse capabilities? (b) How important is support for IP creation, protection, delivery, and reuse? (c) How important is support for externally generated models that conform to OMF and VSI emerging standards?

2.4 Removal of Unused Language Features

As shown in Figure 9(a), when asked which current features in VHDL should be removed to simplify the language, a significant percentage of the respondents indicated "none."

In addition, a majority of the respondents indicated that if a trade-off in compatibility between VHDL 200X and earlier versions of VHDL vs. increased capability AND ease of use were to be made, than compatibility should be the overriding concern. This result, shown in Figure 9(b), clearly indicates that any removal of features from the language should be done carefully and with an eye towards the capability of continued use of legacy code.

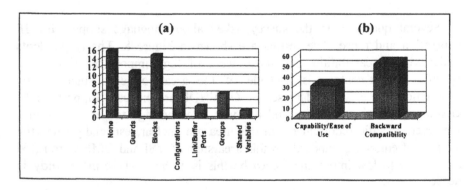

Figure 9. (a) In the interest of simplifying VHDL what language constructs could be eliminated? (b) Should the emphasis in the next revision of VHDL be on language capability and readability/ ease of use vs. backward compatibility with existing version?

2.5   User Interface (Tool) Issues

The survey asked a question about the point in the design process where the users spent a majority of their time and many of the respondents indicated that to be simulation/verification. This response seems to be reinforced by the responses to questions on user interface and tool issues. In this area, the users indicated that accelerated simulation via cycle-based, synthesizable or other subset standardization was of high importance as shown in Figure 10(a).

In addition, they indicated that a standard simulation control language is of high importance as shown in Figure 10(b). Both of these improvements would help reduce the time they need to run the large numbers of simulations (automated via simulation scripts) that are required to verify a design as it is refined to an implementation. The requirement of a standard simulation database format was rated as somewhat less important perhaps indicating that verification data analysis is not currently a problem.

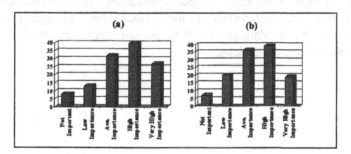

Figure 10. (a) How important is accelerated simulation via cycle-based, synthesizable or other subset standardization? (b) a standard simulation control language?

3.   THE NEXT STEP

The next step in the requirements process is to conduct the second phase of the user survey. This phase will begin by selecting candidate respondents from the first phase of the survey for follow-up interviews. Other key people who have not responded to the survey may also be contacted (e.g. those from the test community). These interviews will be focused on in-depth discussions of the user's needs for the future and how these needs can translate into requirements for VHDL 200X.

Once the second phase of the user survey is completed, the detailed requirements generation process can begin. Detailed requirements for changes or additions to the VHDL language should be identified and documented. These requirements can then be reviewed in detail with the VHDL language development community (the IEEE CS VASG and other working groups within the DASC), and the VHDL tool vendors, to ensure that the recommended requirements have the greatest chance of being addressed by the language update.

The design environment in mid-2000 will be complex, not only because of silicon densities, but also because of the multiple means of creating models, at many levels of the design hierarchy, during the system design process. Various overlapping communities are investigating VHDL and Verilog enhancements, SLDL, the use of C++ and Java for hardware/software design, and special hardware as an engine for system design (coupled with its own design tools). Software models will be very much a part of the overall system model. A seamless path between these various model representations, at the various levels of the design hierarchy, and across any given level of that hierarchy, must be available. Therefore, the extensions to VHDL by 200X must address such features as behavioral synthesis, system integration, object orientation, system-level constraint specification, as well as interfaces to the various other languages that will be in use. Hopefully, the data presented herein will help form the basis of the development of these extensions.

Application of VHDL Features for Optimisation of Functional Validation Quality Measurement

Celia López[1]: Teresa Riesgo[2]: Yago Torroja[2]: Javier Uceda[2]: Luis Entrena[1]
[1]*Area de Tecnología Electrónica. Universidad Carlos III de Madrid. Spain:* [2]*División de Ingeniería Electrónica. Universidad Politécnica de Madrid. Spain*

Key words: Functional validation, Quality measurement, CAD tools and Error model.

Abstract: Functional validation plays an important role for detecting design errors in the
             design process of digital integrated circuits. Early error detection improves
             design quality and reduces re-design costs. The measurement of the quality of
             functional validation with respect to the detection of design errors will help
             designers to enhance their design descriptions. An automatic tool has been
             proposed and developed in order to measure functional validation quality for
             design descriptions in VHDL at the RT level. The **error model** concept and
             the **error simulation** technique have been applied to perform this quality
             measurement. Error simulation is the most critical task in terms of system
             resources needed and its automation should be deeply studied. VHDL includes
             features that may help to optimise the tool performance. Three aspects of
             VHDL are analysed and compared with respect to the benefits they provide.
             These aspects are: flexibility of data types, configuration possibilities and
             event-driven simulation with respect to procedure activation. The experiments
             performed have validated the adopted techniques and have proven the
             advantages they provide to the optimisation of the automatic tool developed

13

P.J. Ashenden et al. (eds.), System-on-Chip Methodologies & Design Languages, 13–24.
© 2001 *Kluwer Academic Publishers. Printed in the Netherlands.*

1. INTRODUCTION

The design of digital integrated circuits is becoming a fully automated process. This automation may enhance the quality of the final product because no designers' errors will be introduced during final steps of the design cycle. On the other hand, design errors are still being introduced in the first steps, where functionality is regularly being described by means of a HDL (*Hardware Description Language*). If these design errors are not detected early, they may be propagated to the final circuit, due to design automation. Therefore, it is necessary to detect as many design errors as possible in the first steps of the design cycle.

Functional validation based on simulation has been a traditional way of checking the absence of design errors. If this validation is good enough, it will detect a large number of design errors. Following the initiative of generating tools to check the quality of all elements involved in design process [1], we have proposed and developed an automatic tool to measure the quality of functional validation with respect to the detection of design errors. This tool will be applied in the first steps of design cycle, to designs described in VHDL at the Register Transfer Level (RTL) of abstraction. In the first stages of the design process it is cheaper to eliminate design errors than in lower abstraction levels, where functionality is not explicit.

Validation Quality Tool (VQT) measures the quality of functional validation by detecting those parts of the VHDL description that are not being checked enough with a given test bench. Designers could enhance original test bench in order to check all functionality described.

Related to the quality in functional validation, several authors have studied the problem. From a theoretic point of view, Aas [2] and Reventlow [3] have analysed the origin of design errors and the way these errors could be reduced. These studies do not provide a practical method to generate or to evaluate the quality of a functional test bench that detects design errors. In a more practical approach, some authors have developed methods for the generation of high-quality functional test benches. The methods proposed are developed in different abstraction levels. Szygenda [4] deals with logic level descriptions, while Allingham [5] works at the algorithmic level descriptions (using a specific language based on C: Wisil).

Most of the initiatives dealing with the quality of functional test benches based on the detection of design errors have intended to propose a method to generate good test benches. Only a few works have let this task to designers and have tried to obtain a method to measure the quality of an existing test bench. At the RT level and with VHDL descriptions, Riesgo [6] has proposed a practical method to measure quality of functional test benches. First, after a study of the problem she has defined an error model for VHDL

RTL descriptions. This error model represents those design errors that are more likely introduced during the description of the design.

We apply this concept of error model and the error simulation technique to measure functional validation quality. Error simulation executes functional validation and compares the behaviour of the original design description against another description with errors inserted. When differences between behaviours of both descriptions are detected, functional validation is said to detect design errors. We represent the quality of the functional validation by the ratio between detected and inserted errors, named error coverage, [6]. VQT implements these three concepts: error models, error simulation and error coverage, and applies them in an automatic way in order to make the method more general and applicable. Error simulation is the most critical task with respect to system resources needed: memory size and execution time involved. All techniques included in the method have been studied with respect to the system resources optimisation. The use of VHDL descriptions is not accidental; this language includes several advantages that will help to optimise the tool.

Among the advantages offered by VHDL when dealing with design descriptions at the RTL level, we have selected data types and overloaded procedure possibilities. In this paper, these two aspects of VHDL will be analysed and compared with respect to the benefits they provide to the Validation Quality Measurement Tool. Those benefits are evaluated with the presentation of experimental results that point out a good enhancement in system resources employed by the developed tool. This chapter is organised as follows. First section deals with theoretic concepts of error model and error simulation. Afterwards second section shows the proposed method and the developed tool. Third section analyses different techniques considered to optimise the tool, special features of VHDL language are pointed out in this section. Fourth section shows some experimental results and analyses the advantages of applying the proposed techniques. Finally, fifth section presents conclusions about the work developed.

2. Error Model and Error Simulation

The proposed method is based on the assumption that it is possible to measure the quality of functional validation with an error model. Two aspects are required to obtain good measurements. The first one is the accuracy of the error model, as it should represent design errors introduced by designers as accurately as possible. The second one is the quality of design descriptions; we will apply the hypothesis called "efficient designer"

[6], which assumes that a small number of design errors have been introduced. Taking into account these two aspects, it is possible to measure the quality of a functional test bench with an error model, considering that if the test bench detects a large number of errors from the model, it would detect also a large number of real design errors.

The input VHDL descriptions must be synthesisable and, therefore, the error model has been adapted to the VHDL Standard for Synthesis [7], focusing mainly on the hardware inferred from VHDL elements. Besides, in order to develop a "formal" method for measuring the quality of functional validation, an exhaustive study has been executed on error model and VHDL elements in order to establish relationships among them in any case.

Element	*Stuck Error*	*Switch Error*	*Dead Error*
Objects	Targets of assignments are taking values belonging to its range	Targets of assignments are changed by another object of the same type	Not applicable
Expressions	Expressions are taking values belonging to their range	Elements in expressions are changed by another of the same type	Not applicable
Statements	Not applicable	Not applicable	Assignments, single or grouped, are not executed

Table 1. Stuck, Switch and Dead Errors

Adopted error model applies to the three main elements in VHDL: data objects, expressions and statements. Error classes are Stuck, Switch and Dead. *Stuck errors* fix values to an affected element, *Switch errors* interchange elements of the same type and, *Dead errors* disable the execution of the involved element. In Table 1 the error model is summarised, further detail may be found in [8], [9], [10] and [11].

3. Quality Measurement Method

As stated in the introduction, we obtain **quality measurement** for functional validation by applying the error model to the original design description. Functional test bench is then simulated in order to check if it is able to detect inserted errors. We have called this simulation **Error simulation** as it compares the behaviour of original description against description with errors inserted. Error insertion is done in single components within design hierarchy, and behaviour is associated to the outputs of the component under validation. The ratio between detected and inserted errors is the **Error coverage**, which represents the quality of checked functional validation. The method has been divided into some steps in order to make it more applicable. Also, these steps have been automated to help designers and project managers to apply the method to their designs. The initial steps are related to the generation of "erroneous" descriptions to be simulated together with original code. In Figure 1 these steps, error list generation and error insertion, are shown. The following steps are related to error simulation and results analysis, Figure 2 shows their application.

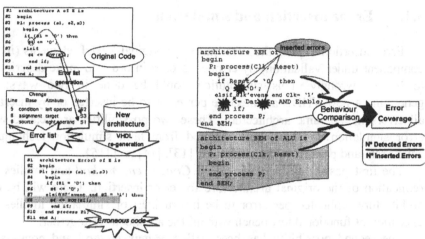

Figure 1. Error list generation and insertion   *Figure 2. Error simulation*

Automation has been enhanced by including a graphical interface, [1], that enables the selection of some parameters, such as types of error inserted, comparison frequency, level of hierarchy, etc. Finally, results could be analysed, not only the quality measurement, but also the non-detected errors that could be seen directly in VHDL original descriptions. Further information about developed automatic tool for the application of quality measurement method could be found in [12].

4. Optimisation techniques from VHDL

As it has been previously said, we perform error insertion on single components within the design hierarchy. Error simulation executes a simulation of the original component together with copies of this component that contain errors inserted (taken from the error model). Application of inputs and comparison of outputs from all components at the same time is an important point in the automation of the method. Execution time requirements will influence greatly in the applicability of both the tool and the method.

VHDL language could help in the reduction of system resources needed. Not only error insertion but also error simulation will be enhanced with the application of specific features of VHDL. Three aspects are presented in this work with respect to automation optimisation: Error insertion and simulation, Error detection by behaviour comparison and Error detection by information collection.

4.1. Error insertion and simulation

Error insertion implies the substitution of original VHDL elements (in the component under validation) by erroneous ones in order to make description to behave wrongly. This substitution could be done in two ways: by generating a new VHDL architecture per error to be inserted or by inserting all errors in the same architecture. These two possibilities are related with error simulation and have been adapted from fault simulation techniques: concurrent and parallel fault simulations [13], [14] and [15].

The first possibility has been called *Concurrent method* and implies the replication of the original architecture for error insertion. There will be one architecture replicated per error to be inserted. Error simulation implies the execution of functional test bench with all the architectures together.

The second possibility has been called *Matrix method* and consists of extending data types and operators in one dimension. Erroneous behaviours are located in this extended dimension (zero element applies for original description) [12].

These two alternatives have been analysed and evaluated. Results have pointed out that no one is better than the other with respect to system resources implied. Nevertheless, interesting conclusions have been extracted. Due to the advantages obtained in execution time, Concurrent method was selected. Next section will detail experimental results.

4.2. Error detection: behaviour comparison

Once error insertion has been solved, other techniques have been studied in order to optimise error simulation. A key problem was the comparison procedure. Design behaviour is mostly linked to the outputs of the studied component, so the output values of the original component are compared against the output values of the erroneous component. This comparison has relative importance in system resources required.

Firstly, the tool developed is aimed to be automatic and should deal with any block interface and, secondly, system resources utilisation should be as low as possible. Three possibilities are considered with respect to output comparison. These possibilities are related to VHDL simulation, so the differences are in process or procedures activation and size of memory needed.

In the first solution considered, all equivalent outputs of original and erroneous components are compared in the same VHDL procedure, this technique has been called *Horizontal comparison*. In the second solution, all outputs of the original component are compared with all outputs of one erroneous component in a specific procedure, this technique has been called *Vertical comparison*. Finally, in the third solution every output of original component is compared with every equivalent output of every erroneous component in a single procedure, this method has been called *Atomic comparison*. In Figure 3 these three solutions are shown graphically, and in Table 2 the differences between the three possibilities are pointed out.

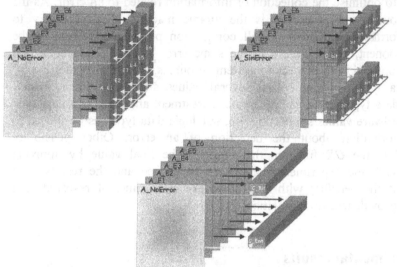

Figure 3. Vertical, Horizontal and Atomic comparison techniques

Several experiments have been performed in order to choose the best technique and the experimental results have concluded that atomic comparison is the best method with respect to system requirements and error simulation needs (execution time and required memory blocks). These experiments are detailed in next section.

Technique	Number of comparison procedures	Activation of procedures	Automation requirements
VERTICAL	Error number	While error is still not detected and outputs compared present events	Necessary customised comparison procedures
HORIZONTAL	Outputs number	While outputs compared present events	Generic procedures possible
ATOMIC	Outputs * Errors	While error is still not	Generic procedures possible

Technique	Number of comparison procedures	Activation of procedures	Automation requirements
		detected and outputs compared present events	

Table 2. Comparison techniques in error simulation

4.3. Error detection: information collection

After choosing **Concurrent method** for error insertion and simulation and **Atomic comparison** for detecting behavioural differences, it is also necessary to optimise the collection of information related to an error. As the selected comparison procedure is the atomic method, it is necessary to collect information provided for all comparison procedures related to the same component, and therefore to the same error inserted. This collection has to be optimal with respect to system resources. VHDL provides a very useful data type that can collect several values and process final result. Resolved data types allow multiple values treatment and *std_logic* data type enables hardware values supporting. So, std_logic data type has been used to collect information about the detection of an error. Other possibility considered is the *OR* function that resolves the final value by applying logical OR. Some experiments have been performed and the results have pointed out the benefits, with respect to execution time, of resolved data types will provide to error simulation.

5. Experimental results

In order to analyse the benefits of applying these techniques to error simulation, some experiments have been performed. The analysed parameters are the execution time as well as the memory used. All the experiments have been executed in Sun Ultra-SPARC machines with Solaris 4.0 operating system. The designs used are academic and industrial designs and comprise not only single blocks but also hierarchical designs. This set includes different functionalities such as bi-dimensional digital filter or arithmetic-logic unit. Academic descriptions are simple designs that could be validated very easily; test bench generation does not present any problem. On the other hand, industrial designs have a medium-high complexity that needs a complete test bench for functional validation.

In Table 3 the main characteristics of all designs involved in experiments are shown.

DESIGN		# VHDL LINES	# STATE MENTS	# OUTPUTS	#IN PUTS	#FLIP FLOPS
ACADEMIC	ALU	109	56	9	19	0
	ALU16	198	86	17	35	0
	FSM4	70	39	3	4	2
	TRANS	60	29	2	11	19
	RECEV	65	32	9	4	21
	SHREG	41	14	8	13	8
	SHR2	31	9	2	4	2
	CONT16	43	15	17	3	17
	MULTIS	81	40	17	21	48
INDUSTRIAL	ASCAT	6.477	800	85	62	340
	OBL	15.000	5.000	35	27	224
	ULTRATEC	8.109	2.100	59	51	951

Table 3

5.1. Concurrent and Matrix methods

To check the best technique for error simulation, concurrent or matrix method, *ALU* design has been selected and the Switch Error type has been inserted.

In this experiment the studied parameters are not only execution time and memory blocks, but also VHDL code lines of erroneous description generated. The obtained results have proven the concurrent method to be better with respect to execution time and worse with respect to VHDL code lines. The memory blocks employed in both methods are in the same order, Figure 4. Although matrix method could seem better a priori, with respect to execution time and memory employed, the experimental results deny this hypothesis. This is due to the use of very complex functions for assigning errors to extended type objects. Some comparison has been also done with respect to the type of computer used. Different SPARC systems have been used and obtained results can be seen in Figure 5. The better the work-station is, the less the difference between both methods is.

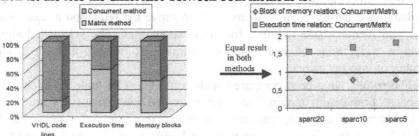

Figure 4 and Figure 5. Results of Concurrent and matrix methods

5.2.   Horizontal, Vertical and Atomic comparison

Among the three comparison techniques proposed, vertical comparison has not been considered due to the problems this solution will imply in the method automation. Atomic and horizontal comparison techniques have been studied in application to academic designs with a wide range of outputs number and VHDL code lines. In Table 4 execution times relation between horizontal and atomic comparison are shown. Figure 6 and Figure 7 show this relation graphically, with respect to the number of outputs and the error coverage obtained.

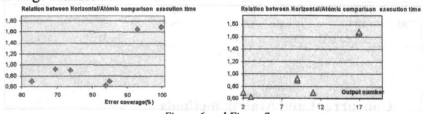

Figure 6 and Figure 7

Design	Number of outputs	# Errors	# VHDL lines	Time relation (t_{horiz}/t_{atom})	Coverage (%)
FSM4	3	25	70	0,62	84
TRANS	2	20	60	0,69	85
FIR2CORE	11	35	188	0,70	63
RECEV	9	23	65	0,90	74
ALU	9	23	109	0,93	70
MULTIS	17	29	81	1,66	93
CONT16	17	11	43	1,69	100

Table 4

5.3.   Information collection

VHDL data resolved type, *std_logic*, has been applied to collect error detection information. Several experiments have been performed on academic and industrial designs, in order to prove the reduction in execution time and memory size required for error simulation. The selected designs cover a wide range of functionalities, number of outputs and types and number of inserted errors.

The results have been compared with respect to OR logic function. The analysis of the results obtained has shown that resolved data types are better in collecting information, although the advantages are not very large. In Table 4, execution times are shown and in Figure 8 and Figure 9 the relation between both functions (OR and resolved) are traced against number of outputs and error coverage. It can be observed that the number of outputs does not influence the ratio, meanwhile error coverage does.

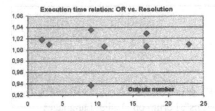 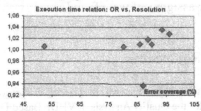

Figure 8 and Figure 9. Execution time relation vs. outputs number and vs. Error coverage

Module	#Outputs	Error Coverage (%)			# Errors			Execution time relation (t_{OR}/t_{Resol})		
		DE	ST	SW	DE	ST	SW	DE	ST	SW
ALU	9	70	100	100	23	50	36	1,053	1	1
FIR2CORE	11	63	87	81	26	59	83	1,005	0,947	1
FSM4	3	84	98	82	25	59	38	1,013	0,857	1,200
MULTIS	17	93	98	95	30	82	56	1,009	1,063	1,091
RECEV	9	74	87	100	24	52	25	0,943	1	0,750
TCM	17	30	50	64	96	183	221	0,999	0,997	1,019
TM	23	81	93	84	76	174	398	0,988	1,014	1,012
TRANS	2	85	85	100	21	45	22	1,066	0,833	1

Table 5

6. Conclusions

In this chapter, a **method for measuring the quality of functional validation** of digital integrated circuits has been proposed. This method has been developed and implemented in an **automatic tool** that helps designers and project managers to measure the quality of their functional test benches in early stages of the design process. The proposed method applies to VHDL descriptions at the Register Transfer Level of abstraction. An automatic tool development has been performed looking for optimum techniques that could enlighten system resources needs. The most time consuming step in the method application is the error simulation, and therefore some techniques have been adapted from fault simulation, that has an extended background about optimisation in simulation. On the other hand, **VHDL may benefit error simulation** also, as it is a *hardware description language* with a large simulation basis. Analysing VHDL structures and features, it has been possible to enhance some aspects of error simulation. In this chapter VHDL advantages in error insertion and simulation and error detection (comparison and collection of information) have been studied and validated with several experiments. Further experimentation could be done in order to apply more techniques, especially with respect to Matrix method, for error insertion and simulation. This technique will be a great solution if optimised procedures of error assignment were found.

7. References

[1] Y. Torroja, C. López, M. García, T. Riesgo, E. de la Torre, J. Uceda, "ARDID: A Tool for the Quality Analysis of VHDL Based Designs", Forum on Design Languages (FDL'99), Lyon (Francia), 1999.

[2] E.J. Aas, T. Steen, K. Klingsheim "Quantifying Design Quality Through Design Experiments", IEEE Design & Test of Computers, Spring 1994. pp. 27-37

[3] C.v. Reventlow "Comparing Quality Assurance Methods and the Resulting Design Strategies: Experiences from Complex Designs", Journal On Electronic Testing: Theory and Applications, 5, 269-272 (1994), pp. 143-146

[4] S. Kang, S.A. Szygenda "Design Validation Comparing Theoretical and Empirical Results of Design Error Modeling". IEEE Design & Test of Computers, Spring 1994. pp. 18-26

[5] D. Allingham, P. Bashford, M. Peters, D. Vendl "Design Test®: A solution to the problems of ASIC verification", International Test Conference, 1989. Paper 39.1, pp. 893-902

[6] T. Riesgo, Ph.D.: "Modelado de fallos y estimación de los procesos de validación funcional de circuitos digitales descritos en VHDL sintetizable", Department of Automation, Electronic Engineering and Industrial Computing. Universidad Politécnica de Madrid, 1996

[7] VHDL Synthesis Interoperability Working Group of the Design Automation Standards Committee, IEEE P1076.6, "Standard For VHDL Register Transfer Level Synthesis", 17 marzo 1998.

[8] T. Riesgo, Y. Torroja, C. López, J. Uceda "Estimation of the Quality of Design Validation Experiments Based on Error Models" VHDL Users Forum in Europe 1997, Toledo (Spain), pp 83-92

[9] T. Riesgo, Y. Torroja, E. de la Torre, J. Uceda "Quality Estimation and Functional Procedures Based on Fault and Error Models", Design Automation and Test in Europe, Paris (France) 1998, pp. 955-956

[10] C. López, T. Riesgo, Y. Torroja, E. de la Torre, J. Uceda "An Error Simulator To Estimate The Quality Of Design Validation Experiments", Forum on Design Languages 1998, Laussane (Switzerland)

[11] C. López, T. Riesgo, Y. Torroja, E. de la Torre, J. Uceda, "A method to perform error simulation in VHDL", Design of Circuits and Integrated Systems Conference 1998. Madrid (Spain), pp 495-500

[12] C. López, Ph.D.: "Estrategia de validación funcional de circuitos digitales descritos en VHDL sintetizable basada en modelos de error", Department of Automation, Electronic Engineering and Industrial Computing. Universidad Politécnica de Madrid, 1999

[13] M. Abramovici, M.A.Breuer, A.D. Friedman, "Digital System Testing and Testable Design" IEEE Press, 1990

[14] G.Russell, I.L. Sayers "Advance Simulation and test methodologies for VLSI Design" Van Nostrand Reinhold International, 1989

[15] E.G. Ulrich, E.T. Baker "Concurrent Simulation of nearly identical digital networks", Computer, April 1974, pp 39-44

AN OBJECT-ORIENTED COMPONENT MODEL USING STANDARD VHDL FOR MIXED ABSTRACTION LEVEL DESIGN

C. Hansen, O. Bringmann, W. Rosenstiel

FZI Karlsruhe
Department Microelectronics System Design
Karlsruhe, Germany

1.1 INTRODUCTION

In complex system design, it is often desirable to start the system specification at higher levels of abstractions, e.g. at the algorithmic level. The necessary refinements are then produced by commercial or academic high-level synthesis systems. More and more often, the integration of user-defined RT components in the algorithmic specification plays an important role. First, some functional and timing behavior can only be implemented at the RT level, e.g. interrupt handling, and interface components. Second, several RT components may already exist and are appropriate for reuse. Third, the re-implementation of VHDL models emulating this behavior at the algorithmic level is expensive and time-consuming. Finally, several synthesis, simulation, and test environments exist which can be used for descriptions at different abstraction levels. Therefore, this paper addresses the prob-

25

P.J. Ashenden et al. (eds.), System-on-Chip Methodologies & Design Languages, 25–36.
© 2001 *Kluwer Academic Publishers. Printed in the Netherlands.*

lem of mixed abstraction level specifications for simulation and behavioral synthesis using object oriented component models. For this, the VHDL standard [IEEE93] without any extensions is used and the usual simulation and synthesis systems can be applied. The communication between algorithmic descriptions and VHDL components at the same or at lower levels is executed using VHDL procedures. To reduce the design time required for the insertion of these procedures in the algorithmic specification, a preprocessor has been developed. The preprocessor allows the procedures to be applied without any extensive declarations of the corresponding RT components. The implementation of procedures emulating the component behavior at the algorithmic level is also possible.

The different timing aspects at algorithmic, RT, or lower levels are encapsulated in the provided procedures. As a result, the algorithmic specification does not have to contain any timing aspects to communicate with components at lower abstraction levels. Finally, an appropriate stimuli set can easily be included testing the integrated VHDL component or the procedure together with the component.

1.1.1 Related Work

A closer investigation of existing approaches shows that two different methods can be identified. First, there exist several object-oriented (OO) approaches extending VHDL with the corresponding OO language constructs (OO-VHDL) [Ash98] [Jerr97] [Rad97] [Rad99] [Schu95] [Swam95]. Using one of these approaches, a mixed abstraction level simulation and synthesis can easily be executed. However, an extension of the standardized language VHDL is necessary. Consequently, commercial simulators and synthesis systems, as well as most of the scientific tools, can not be applied.

This paper is organized as follows: Section 1.2 and Section 1.3 give a general overview of our OO VHDL component model for mixed abstraction level simulation and synthesis. In Section 1.4, the creation of a OO library and the integration into our high-level synthesis system CADDY-II are described. Section 1.5 and Section 1.6 present further synthesis and simulation aspects concerning our approach. This paper concludes with a summary in Section 1.7.

1.2 AN OBJECT-ORIENTED VHDL COMPONENT MODEL FOR SIMULATION AND SYNTHESIS

High-level synthesis systems transform an abstract behavioral specification into a structure of RT components and a finite state machine. Therefore, the main problem for high-level synthesis with mixed abstraction level simulation is how the RT components can be integrated in the behavioral specification. In principle, there exist two possibilities: a procedure is implemented emulating the behavior of the RT component (*emulation procedure*), or the RT component is instantiated in the specification. Using the first method, not every possible design task can be solved. As already mentioned, certain functional and timing aspects can only be implemented at the RT level.

Algorithmic specifications written in VHDL are embedded in a VHDL process. Unfortunately, a component instantiation is not allowed outside of a VHDL process. Thus, the second method can only applied in the form of a structural solution. As a result, the timing signals needed at the RT level, e.g. clock and reset signals, must also be used in the algorithmic specification. One way to avoid this problem is to use procedures implementing the communication with the RT components (*access procedures*). This method has several advantages. Procedures can be used in VHDL processes. The RT timing as well as the communication protocols can be encapsulated in the access procedures. Every procedure has a corresponding RT component forming a unit for simulation (communication aspect) and synthesis (correspondence aspect) (Figure 1.1). To implement this unit in VHDL, in our approach, we use a "frame component" concept. Every frame component contains the RT component to be reused and all corresponding procedures. In object-oriented view, the frame component represents a class, and the procedures the corresponding methods. Once, a RT component are instantiated by the corresponding procedure, then the RT component represents the object.

The synthesis process can use the information about the relation between procedures and RT components expressed by the frame component for an optimized integration of the RT components. Further, a mixed abstraction level simulation is possible. The frame component is implemented as a usual VHDL component. Thus, a library can easily be created using the VHDL library concept.

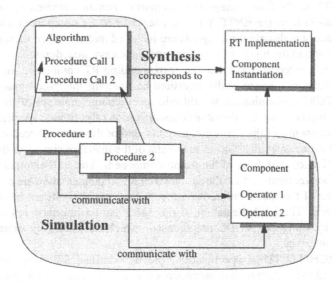

Figure 1.1 Access procedures and their RT components

1.3 THE FRAME COMPONENT CONCEPT

As mentioned above, the basic unit of our approach is the frame component. The frame component is a standard VHDL component, and contains one user-defined RT component and its corresponding procedures. Therefore, the frame component corresponds to a class, the procedures to its methods, and the RT component represents the object after instantiation in the algorithmic specification. Their relationships is depicted in Figure 1.2. Note that only an object-oriented specification style for RT components is provided but no inheritance and polymorphism are supported.

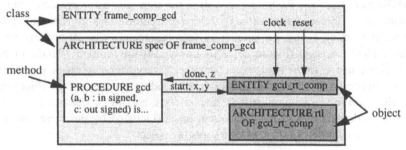

Figure 1.2 Example of a frame component

For every RT component, a frame component has to be defined once. Each procedure represents an operation which can be performed by a RT component. The communication between an access procedure and the corresponding component is performed by global signals defined in the declarative part of the frame component.

The ENTITY of the frame component consists of generic parameters and interface signals taken from the ENTITY of the user-defined RT component. Concerning the interface signals only these signals are included that are not contained in the parameter list and in the body of the corresponding access procedures.

Using access procedures, and thus the corresponding RT component, in a algorithmic specification, it has to be distinguished between the interface signals of the ARCHITECTURE containing the algorithmic specification (main specification), of the processes representing the algorithmic specification (algorithmic specification), of the RT components, of the frame components, and of the access procedures. The interface of the main specification has to consist of the interface signals of the algorithmic specification and those of the frame component. Thus, all external signals required for the algorithm and for the synthesized RT implementation are contained in the interface of the specification. As a result, the interface will not be changed during synthesis. The implementation of the specification interface can be done manually or by using our VHDL preprocessor which automatically extends the specification interface.

In the ARCHITECTURE specification, the user-defined RT components are instantiated and the corresponding procedures are implemented. These procedures can be access procedures or emulation procedures. For simulation, the implementation of a frame component for emulation procedures are not absolutely necessary (Section 1.6). The frame component are used here to give the synthesis system the necessary mapping information, and to increase the optimization potential (Section 1.5). In contrast, for access procedures, the frame component is required for simulation as well as for synthesis. An important advantage of this approach is that both kinds of procedures allow the implementation of user-defined components at any arbitrary abstraction level supported by VHDL.

However, to implement an access procedure, the declarative part of the ARCHITECTURE of the frame component has to contain the declaration of the user-defined component and all signals required for the communication between the procedure and the component. In the implementation part, the component is instantiated, and the access procedures are implemented in the declarative part of a PROCESS. The implementation part of the PROCESS can particularly be used for specifying simulation vectors in order to test the component including the corresponding procedures.

1.4 THE OBJECT-ORIENTED COMPONENT LIBRARY

An important advantage of the frame component concept is that a reusable object-oriented component library can easily be implemented using the VHDL library concept. For every frame component a corresponding component declaration can be inserted in a VHDL package. This allows an easy extension of the component library.

Furthermore, this package or library can be used by any VHDL synthesis system. In our approach, a reuse component library is applied by our high-level synthesis system CADDY-II (Figure 1.3) [Gut91] [Gut94].

A VHDL preprocessor is used to execute the VHDL specific tasks. The preprocessor consists of one front-end, the VHDL parser, and two back-ends, one for synthesis specific tasks (*synthesis back-end*) and the second for simulation specific tasks (*simulation back-end*). The tasks of the simulation back-end are described in Section 1.6.

The synthesis back-end translates the VHDL specification into a specification described in an internal intermediate format (Figure 1.3 (a)). This format is used by the CADDY-II synthesis system representing a language independent input description. Further, the synthesis back-end is used to transform the VHDL specification of the reuse component library into the same internal intermediate format (Figure 1.3 (b)). This has to be done only if new frame components have been inserted in the library. Finally, CADDY-II can execute the usual high-level synthesis process. During this process, the operations of the algorithm are mapped onto components which provide the corresponding operators (Figure 1.1). Not only user-defined operators, but also several standard operators can be used. The standard operators are defined in the IEEE *std_logic_arithm* package, and therefore, need not to be specified in the object-oriented component library.

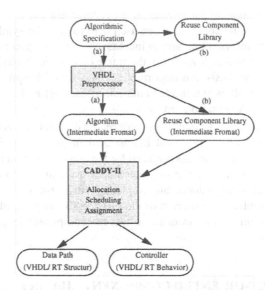

Figure 1.3 Synthesis flow using CADDY-II and the VHDL preprocessor

1.5 FRAME COMPONENT AND SYNTHESIS

For synthesis, the frame component contains a lot of important information. First, the relationship between procedures and the user-defined component are determined simply by defining procedures and the corresponding component in the same frame component. Thus, for every procedure, the synthesis system is given information about the component to be mapped on. Next, synthesis specific information can be specified using VHDL attributes, e.g. number of gates, area, frequency or power. An example for specifying these attributes is given in Listing 1.1. A graphical overview of this example has been given in Figure 1.2.

```
(1)    ENTITY frame_comp_gcd IS
(2)      GENERIC( width1 : integer);
(3)      PORT(clock,reset : IN std_logic);
(4)    END frame_comp_gcd;
(5)
(6)    ARCHITECTURE specification OF frame_comp_gcd IS
(7)      -- architecture declarative part
(8)      COMPONENT gcd_rt_comp
(9)        GENERIC(width1 : integer);
(10)       PORT(clock, reset : IN std_logic;
(11)         start : IN std_logic;
(12)         done : OUT std_logic;
(13)         x,y : IN signed(width1-1 DOWNTO 0);
(14)         z : OUT signed(width1-1 DOWNTO 0));
(15)     END COMPONENT;
```

```
(16)    ATTRIBUTE Gates OF gcd_rt_comp:
           COMPONENT IS 5*width1+4;  -- XC4000 CLBs
(17)    ATTRIBUTE Area OF gcd_rt_comp:
           COMPONENT IS ((5*width1 )/2+4, (5*width1)/2);
(18)    ATTRIBUTE Power OF gcd_rt_comp: COMPONENT IS 1.5*width1;
(19)    ATTRIBUTE Frequency OF gcd_rt_comp: COMPONENT IS 120/width1;
(20)    SIGNAL start, done : std_logic;
(21)    SIGNAL x,y,z : integer;
(22)
(23)    BEGIN
(24)    gcd1 : gcd_rt_comp
(25)       GENERIC MAP (width1);
(26)       PORT MAP (clock, reset, start, done, x,y,z);
(27)    PROCESS
(28)       -- process declarative part
(29)       PROCEDURE gcd(a,b,c : IN signed(width1-1 DOWNTO 0)) IS
(30)          ATTRIBUTE Commutativity OF a : CONSTANT 1;
(31)          ATTRIBUTE Commutativity OF b : CONSTANT 1;
(32)          ATTRIBUTE Commutativity OF c : VARIABLE IS 2;
(33)          ATTRIBUTE InitiationInterval OF ALL : CONSTANT IS 0;
(34)       BEGIN
(35)          specification.start <= '1';
(36)          specification.x <= a;
(37)          specification.y <= b;
(38)          WAIT UNTIL specification done'event
                    AND specification.done - '1';
(39)          specification.start <= '0';
(40)          c := specification.z;
(41)          WAIT FOR 1 ns;
(42)       END gcd;
(43)    BEGIN
(44)       -- stimuli set for access procedures / RT component
(45)    END PROCESS;
(46)    END specification;
```

Listing 1.1 Frame component with synthesis attributes

Besides the attributes *Area, Power* etc. referring to the whole component, procedure specific attributes can also be specified. Using the attribute *Commutativity* it can be specified which input or output signals can be interchanged. The attribute *InitiationInterval* defines a period after which the component can be started again. Furthermore, this attribute is used to distinguish combinational and sequential implementations. If this attribute is set to the value '0', a combinational component is used, setting the attribute to a value greater than '0' causes a sequential component to be used.

The synthesis process can use the information concerning which procedures correspond to which component for an optimized allocation. Procedures that are contained in more than one frame component using the same name and implementing the same functionality offer the same optimization possibilities as standard procedures representing standard operators. In correspondence to the design constraints the synthesis system can now allocate the component with the best area/performance costs given by the attributes. A further possibility to determine the component to be allocated is that the designer explicitly specifies a reference to the corresponding component. This can be done by using the attribute *map_to_operator*.

1.6 FRAME COMPONENT AND SIMULATION

For simulation, the main aspects concerning the integration of user-defined components in an algorithmic specification are: the communication between the access procedures and the RT components (Figure 1.1 (Simulation)), and the implementation of simulation vectors. Note that emulation procedures does not cause any problems, because they can be used directly.

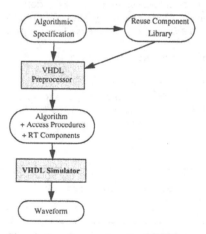

Figure 1.4 Simulation flow using the VHDL preprocessor

As mentioned before, the access procedures and components communicate via global signals with each other. This means that in every specification in which access procedures are used, the declarative part of the ARCHITECTURE has to contain the declaration of the user-defined component as well as all necessary global signals. Further, the implementation part of the ARCHITECTURE has to contain the component instantiation and the implementation of the procedure in the PROCESS declarative part. These hand-written descriptions are very time consuming. Due to the fact that the required information is already given in the corresponding frame component, the same preprocessor used for synthesis but the other back-end is applied to insert all necessary information in the specification. In this case, the back-end for simulation specific tasks (*simulation back-end*) is used. In contrast to the synthesis back-end, the simulation back-end generates a VHDL description instead of the language-independent intermediate format. The design flow containing our VHDL preprocessor is given in Figure 1.4.

Applying this preprocessor, the user has only to define the access procedures and a reference to the frame component to be used for simulation. During synthesis, this reference is ignored to allow optimization possibilities. In Listing 1.2, a short example using an access procedure in a specification is given. The access procedure is called in line 23. The information to be inserted from the preprocessor are listed in the comments.

```
(1)      LIBRARY ieee;
(2)      USE ieee.std_logic_1164.ALL;
(3)      USE ieee.std_logic_arith.ALL;
(4)
(5)      -- pragma DSL USE clib.reuse_comp_library.frame_comp_gcd;
(6)
(7)      ENTITY test IS
(8)        PORT(-- automatic extension of the interface
(9)          in1    : IN  signed(15 DOWNTO 0);
(10)          in2    : IN  signed(15 DOWNTO 0);
(11)          out1   : OUT signed(15 DOWNTO 0));
(12)      END test;
(13)
(14)     ARCHITECTURE example1 OF test IS
(15)       -- automatic insertion of global signals,
(16)       -- and component declarations
(17)     BEGIN
(18)       -- automatic insertion of the component instantiation
(19)       PROCESS
(20)         -- automatic insertion of procedure implementation
(21)         VARIABLE result : signed(15 DOWNTO 0);
(22)       BEGIN
(23)         gcd(in1, in2, result);
(24)         out1 <= result;
(25)       END PROCESS;
(26)     END example1;
```

Listing 1.2 Specification using an access procedure

The component reference is specified using a pragma construct (Listing 1.2 (line 5)). With this construct, the library, the package, and the frame component are defined in which the preprocessor can find the procedure and the corresponding component. For our preprocessor concept, several problems had to be solved:

- The user-defined component may contain generic parameters, e.g. the width of one or several bitvector(s). In this case, the preprocessor extracts the necessary generic parameters from the given parameters of the procedure. Only, if this is impossible, the user has to define the missing generic parameters using a pragma clause.

- The preprocessor generates the needed global signals. In some cases, it is not possible to simply take over the name of the global signals defined in the frame component. Some of these names may already be used in the algorithmic specification. Thus, the preprocessor generates new names for those global signals.

- The correct reference, and the complete logical name of every global signal used in the procedure body are generated by the preprocessor. In the procedure body, several global signals can be used. To avoid possible name conflicts, these signals are used with their complete logical names.

- The preprocessor automatically extends the specification interface. This interface is extended by all present interface signals of the frame component (Section 1.2). The generic parameters of the frame component must not be inserted, because, in the specification, the actual values are used.

- The preprocessor automatically determines the relation between procedure and pragma construct. When several procedures are used in the specification which belong to different frame components, several relation conflicts may occur. E.g., two procedures *pa* and *pb* belong to two frame components *fa* and *fb* and are referenced by two pragma constructs. Now, procedure *pa* is contained in frame component *fa* as well as in frame component *fb*. First, the preprocessor uses the procedure parameter as indication for the correct relation. Assuming that the name of the procedure as well as the parameters are equal, the preprocessor suggests the most probable relation, but the user has to confirm or to correct this suggestion in order to generate a unique assignment for simulation.

The resulting VHDL preprocessor listing can be seen in Listing 1.3. The reference *example1* is used to avoid possible name conflicts.

```
(1)      LIBRARY ieee;
(2)      USE ieee.std_logic_1164.ALL; USE ieee.std_logic_arith.ALL;
(3)
(4)      ENTITY test IS
(5)        PORT(-- automatic extension of the interface
(6)          clock, reset : IN std_logic;
(7)          in1, in2 : IN  signed(15 DOWNTO 0);
(8)          out1  : OUT signed(15 DOWNTO 0));
(9)      END test;
(10)     ARCHITECTURE example1 OF test IS
(11)       -- automatic insertion of the component declaration,
(12)       COMPONENT gcd_rt_comp
(13)         GENERIC(width1 : integer);
(14)         PORT(clock, reset : IN std_logic;
(15)           start : IN std_logic;
(16)           done : OUT std_logic;
(17)           x,y : IN signed(width1-1 DOWNTO 0);
(18)           z : OUT signed(width1-1 DOWNTO 0));
(19)       END COMPONENT;
```

```
(20)      -- automatic insertion of the global signals
(21)      SIGNAL start, done : std_logic;
(22)      SIGNAL x,y,z : signed(15 DOWNTO 0);
(23)   BEGIN
(24)      -- automatic insertion of the component instantiation
(25)      gcd1 : gcd_rt_comp
(26)         GENERIC MAP(16);
(27)         PORT MAP (clock, reset, start, done; x,y,z);
(28)
(29)      PROCESS
(30)         -- automatic insertion of the procedure declaration
(31)         -- and implementation
(32)         PROCEDURE gcd( a, b, c: IN signed(width1-1 DOWNTO 0)) IS
(33)         BEGIN
(34)            example1.start <= '1';
(35)            example1.x <= a;
(36)            example1.y <= b;
(37)            WAIT UNTIL example1.done'event AND example1.done = '1';
(38)            example1.start <= '0';
(39)            c := example1.z;
(40)            WAIT FOR 1 ns;
(41)         END gcd;
(42)         VARIABLE result : signed(15 DOWNTO 0);
(43)      BEGIN
(44)         gcd(in1, in2, result);
(45)         out1 <= result;
(46)      END PROCESS;
(47)   END example1;
```

Listing 1.3 VHDL specification after preprocessing

Until now the communication between access procedure and the corresponding component as well as the resulting preprocessor concept has been discussed. Next, simulation vectors have to be integrated. In our approach, in the frame component, simulation vectors for every procedure and the corresponding component (Listing 1.1 (line 44)) can easily be inserted. The most appropriate place for these vectors is the implementation part of the PROCESS. Here, stimuli, which test the access procedure together with the component, can easily be inserted, as well as stimuli, which only test the component or an emulation procedure. As a result, "standard" simulation vectors are available at once, and they may be an important part in every further reuse and validation process [Han97].

1.7 CONCLUSION

This paper presented a new approach for designing VHDL models for mixed abstraction level simulation and behavioral synthesis. The advantages of the presented approach are: First, the frame component concept offers all necessary properties for mixed abstraction level simulation and synthesis using an object-oriented specification style for RT components. Concerning simulation, the access/emulation procedures and the corresponding user-defined component are contained in a frame component specifying a class in object-oriented design. Thus, our VHDL preprocessor can optimally support the designer applying access procedures in an algorithmic specification, and executing the final simulation. Concerning synthesis, specific attributes can be defined to extend the further optimization potential during the high-level synthesis process. Second, the user-defined components can be implemented at an arbitrary abstraction level supported by VHDL. Third, a object-oriented component library can easily be implemented using the VHDL library concept. Finally, our approach is based on the IEEE VHDL standard and can therefore easily be used by any simulation and synthesis tool supporting VHDL.

1.8 REFERENCES

[Ash98]   Ashenden, P.J.; Wilsey, P.A.; Martin, D.E.: *"SUAVE: Object-Oriented and Genericity Extensions to VHDL for High-Level Modeling"*, Proceedings of FDL, 1998.

[Bring97]   Bringmann, O.; Rosenstiel, W.: *"Cross-Level Hierarchical High-Level Synthesis"*, Proceedings of D.A.T.E., 1997.

[Gut91]   Gutberlet, P.; Krämer, H.; Rosenstiel, W.: *"CASCH - a Scheduling Algorithm for High Level -Synthesis"*, Proceedings of the EDAC, February 1991.

[Gut94]   Gutberlet, P.; Rosenstiel, W.: *"Timing Preserving Interface Transformations for the Synthesis of Behavioural VHDL"*, Proceedings of EURO-DAC, September 1994.

[Han97]   Hansen, C.; Kunzmann, A.; Rosenstiel, W.: *"Verification by Simulation Comparison Using Interface Synthesis"*, Proceedings of D.A.T.E., 1997.

[IEEE93]   IEEE: *"VHDL Language Reference Manual"*, ANSI/IEEE Standard 1076-1993, June 1993.

[Jerr97]   Jerrraya, A.A.; Ding, H.; Kission, P.; Rahmouni, M.: *"Behavioral Synthesis and Component Reuse with VHDL"*, Kluwer Academic Publishers, 1997.

[Rad97]   Radetzki, M.; Putzke-Röming, W.; Nebel, W.: *"Objective VHDL: The Object-Oriented Approach to Hardware Reuse"*, in: Roger, J.-Y.; Stanford-Smith, B.; Kidd, P.T. (eds.): Advances in Information Technologies: The Business Challenge. IOS Press, Amsterdam, 1998. Presented at EMMSEC'97, Florence, Italy, 1997.

[Rad99]   M. Radetzki: *"Overview of Objective VHDL Language Features"*; Proceedings of FDL, 1999.

[Schu95]   Schumacher, G.; Nebel, W.: *"Inheritance Concept for Signals in Object-Oriented Extensions to VHDL"*, Proceedings of European Design Automation Conference, 1995.

[Swam95]   Swamy, S.; Molin, A.; Covnot, B.: *"OO-VHDL: Object-oriented extensions to VHDL"*, IEEE Computer, vol. 28, no.10, October 1995.

A VHDL-Centric Mixed-Language Simulation Environment

A.Windisch[1,3], D.Monjau[1], T.Schneider[2,3], J.Mades[2,3], M.Glesner[2], and W.Ecker[3]

[1] *TU Chemnitz, Department of Computer Science, 09107 Chemnitz, Germany*
[2] *TU Darmstadt, Institute of Microelectronic Systems, Karlstr. 5, 64283 Darmstadt, Germany*
[3] *Infineon Technologies AG, CPD DAT ADM, Otto-Hahn-Ring 6, 81739 Muenchen, Germany*

Keywords: Mixed-Language, Multi-Language, Verification, Simulation, VHDL, JAVA, C++, Co-Simulation

Abstract: This paper presents a mixed-language simulation environment for the languages VHDL, JAVA, and C++. The environment is implemented in JAVA and based upon a previously developed VHDL design environment consisting of a compiler, an elaborator, and a simulator. The latter was extended by open object-oriented JAVA and C++ interfaces towards mixed-language simulation capabilities. Clearly, this approach lends itself to a VHDL-centric modelling style. However, it also results in a well defined overall simulation semantics based on the proven semantic principles of VHDL. Moreover, the object-oriented JAVA and C++ interfaces enforce a much better language modelling style than traditional callback-based procedural language interfaces.

1.   INTRODUCTION

New paradigm shifts in hardware (HW) design, such as System-On-Chip (SoC) and core-based design, require a higher level of HW description and modelling. For the next generation of HW systems traditional designs at register-transfer and behavioural level will not be sufficient any more. Instead future HW systems will be described at system level by a set of interacting HW cores. Thereby, different cores will be implemented on different levels of abstraction using different hardware description languages (HDL) or programming languages (PL).

37

P.J. Ashenden et al. (eds.), System-on-Chip Methodologies & Design Languages, 37–46.
© 2001 *Kluwer Academic Publishers. Printed in the Netherlands.*

Increasing system complexity/size and decreasing Time-To-Market will further drive the need for mixed-language simulation systems. They will be the key to rapid prototyping and interactive simulation by combining HDL semantics with high-level PL semantics. Thus, these new simulation systems will combine the strengths of HDLs with the advantages of PLs, such as graphical user interfaces, database access, and networking. This will enable mixed-language simulation systems to play a dominating role in the future system-level design methodology, in particular in the fields of product definition, interactive testing, and performance analysis and evaluation.

In this paper we present our ongoing work on the implementation of a mixed-language simulation system. The information is structured as follows: The next chapter briefly describes related work and activities in the field of mixed language simulation and design. Afterwards, chapter 3 explains in detail the system architecture and chapter 4 presents an example of a mixed-language design. Finally, chapter 5 summarises the work before concluding with a brief outlook of future activities.

2. RELATED WORK

The importance of mixed-language simulation was recognised early in the standardization efforts of the System Level Design Language (SLDL) [1]. Efforts were made into two research directions for mixed-language simulation: the development of language based approaches [2] and of computation-model based approaches [3, 4].

The language based approach covers the two distinct concepts of language co-simulation and the simulation of models based on a unified model composition format. Numerous applications exist for both concepts: co-simulation environments, such as KOSIM [5] as well as multi-lingual simulators, such as ModelSim and VSS. These two simulators support VHDL/C simulation by providing a Foreign Language Interface (FLI) [6] and a C Language Interface (CLI) [7], respectively. Moreover, ModelSim supports a mixed VHDL/Verilog simulation [8] as described by the formal inter-operability semantics in [9].

The computation model based approach to mixed-language simulation is supported by tools such as Cadence SPW [10]. This heterogeneous environment combines multiple semantic models, such as synchronous data-flow networks (SDF), dynamic data-flow networks (DDF), discrete-event models, and cycle-based models.

The presented mixed-language simulator solely relies on the hardware description semantics of VHDL [11]. However, this semantics does not provide the expressive power needed for system-level design. Therefore, the mixed-language simulator provides open object-oriented interfaces to JAVA and C++ that allow for an simple integration of high-level components implemented in either language. The suitability of JAVA for system-level design was successfully proven in [12].

3. SYSTEM ARCHITECTURE

The presented mixed-language simulation environment is based upon the already existing Infineon VHDL-AMS Environment [13, 14]. It consists of a VHDL-AMS compiler [15], elaborator [16], and simulator and is entirely implemented in JAVA.

3.1 Overall Structure

The existing VHDL simulator was extended towards mixed-language simulation by adding a JAVA and a C++ interface as shown in figure 1. The resulting new simulator is capable of evaluating components implemented in VHDL, JAVA, and C++.

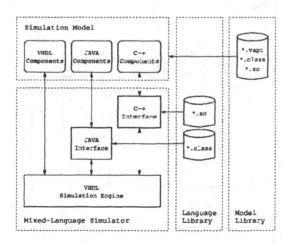

Figure 1. The simulator architecture

The presented approach to mixed-language simulation clearly imposes certain restrictions to the modelling style. First, the top-level simulation model structure must be a VHDL description. This requirement follows directly from the fact that the model must be executable by the VHDL simulation engine. Second, JAVA and C++ components must be modelled as foreign VHDL architectures. This requirement preserves simulation semantics of the overall simulation model. JAVA and C++ are programming

40

languages and therefore only have an execution semantics. Thus, the execution semantics must be extended towards a simulation semantics by adding concepts for simulation time, model structuring, intra-model communication, and model simulation. As previously stated, all model components must be simulated by the VHDL simulation engine. Therefore, we adopt the VHDL time model, the VHDL structure model (component instantiations), the VHDL signal model, and the VHDL process scheduling mechanism, for the aforementioned semantic extensions.

Clearly, the requirements stated above force a VHDL-centric modelling style. However, the advantage of this approach is the resulting well-defined simulation semantics. It is based on the semantic concepts of a single language rather than being a unification of simulation and execution semantics of different languages.

3.2 VHDL Simulation Engine

The VHDL simulation engine consists of a set of model processes, a simulation kernel, and scheduler as shown in figure 2. The set of model processes can be further divided into three distinct subsets denoting VHDL, JAVA, and C++ processes.

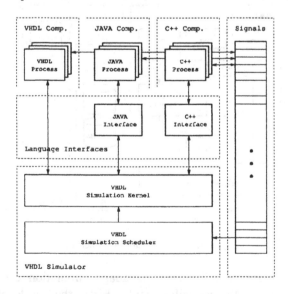

Figure 2. The VHDL simulation engine

Thereby, model processes are logical processes only; they do not represent real physical processes. All model processes are created during design elaboration. For each VHDL process statement and each VHDL concurrent statement a corresponding VHDL process is instantiated in the simulation engine. Elaboration of a foreign JAVA or C++ architecture results in the creation of a JAVA or C++ process instance, respectively. Both JAVA and C++ processes are subsequently called software processes in order to distinguish them from native VHDL processes. All VHDL processes are executed by an interpreter which can access the data-structures built during design compilation and elaboration. This includes full access to the parse trees and the object data structures. As opposed to that, software processes can only access their associated generics, ports, and the actual simulation time. The ports of interface mode IN and INOUT are added statically to the sensitivity set. All software processes are executed as non-postponed processes.

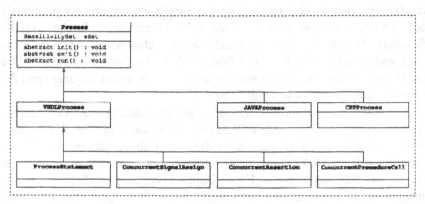

Figure 3. The VHDL process class hierarchy

All model processes are derived from an abstract base class as shown in figure 3. This class defines the simulation protocol specific methods init, run, and exit that are called by the kernel during simulation. The init method is called once for each model process in the initialisation phase. Thereby, VHDL processes are executed until the first wait statement is hit. For software processes this method is user defined. The run method is called by the kernel at every simulation cycle for each process that is sensitive to an active signal. Thereby, the current simulation time is passed to this method as parameter. VHDL processes are executed until a wait statement occurs. For software processes control is passed to their run method until this method exits. It is the responsibility of the user to ensure that this method terminates. Otherwise the simulation gets caught in an infinite cycle. As previously stated, sensitivity sets of software processes are calculated statically at instantiation time. However, sensitivity sets of VHDL processes must be calculated dynamically at each execution of a wait statement. Finally, the exit method is called for each process once at the end of simulation. This does not affect VHDL processes since VHDL does not contain a finalisation semantics. Instead this method is provided for software processes in order to allow cleanup after simulation.

3.3 JAVA Interface

The JAVA interface of the simulator is represented by a JAVA class hierarchy as shown in figure 4. This set of classes can be divided into three distinct subsets denoting different semantic extensions of the JAVA language.

Model structuring and simulation semantics are added to JAVA by class VJArchitecture. JAVA components must be derived from this abstract base class in order to be executed by the simulation engine. As previously stated, JAVA components must always be bound to a foreign VHDL architecture. In the presented approach this is achieved by annotating the foreign architecture with the predefined VHDL attribute FOREIGN. The value of this attribute is used to specify both the component language and the component itself as shown in figure 5 below.

Intra- model communication semantics are provided by the JAVA classes VJInterfaceObject and VJValue. Subclasses of the first denote generic constants and port signals. Instances of these classes are always associated with JAVA components and represent the communication interface of that component. Each instance of an interface object is connected with a corresponding simulator-specific object instance. All communication between the simulator and the associated JAVA component is performed across this two-level interface. The additional level of indirection hides simulator specific implementation details from the user. Furthermore, it results in the implementation of modular, re-usable JAVA components that may be run on different simulators providing the same language interface.

Interface objects provide read and write methods for information passing between the simulator and the associated JAVA component. Unlike [6, 7] we do not map VHDL types to native types since this approach does not allow the representation of composite types. Instead, the type system is modelled implicitly by a hierarchy of value classes derived from the abstract base class VJValue. Instances of these classes may be combined to form composite value instances of arbitrary size. Read and write operations on interface objects are based upon this value instance passing mechanism thus providing a generic and uniform way for information passing between simulator and JAVA component.

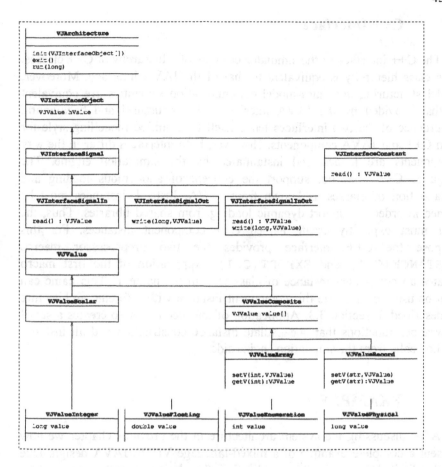

Figure 4. The JAVA interface class hierarchy

```
<foreign_attribute_specification> ::-
    attribute FOREIGN of
    <architecture_name>
    : architecture is
    <foreign_attribute_value> ;

<foreign_attribute_value> ::=
    "JAVA:<class_name>"
  | "CPP:<library_name>:<component_name>"


Component Binding Scheme
```

```
entity E1 is
    -- generic list
    -- port list
    ...
end entity E1;

architecture A1 of E1 is
    attribute FOREIGN of A1:architecture is "JAVA:core.c166";
begin
    -- empty architecture body
end architecture A1;

VHDL Example Code
```

Figure 5. JAVA and C++ component binding

3.4   C++ Interface

The C++ interface of the simulator consists of a hierarchy of C++ classes. This class hierarchy is equivalent to that of the JAVA interface. Moreover, model structuring and intra-model communication semantics are equivalent to that provided by the JAVA interface. This structural and behavioural congruence of the two interfaces lends itself to a unified modelling style for both C++ and JAVA components. However, both interfaces differ in the way components are loaded and instantiated by the simulation engine. The language C++ does not support the concept of anonymous loading and instantiation of classes as known from JAVA. Instead, instances must be named in order to support dynamic loading from shared libraries. Thus, the user must explicitly create named C++ component instances. For this purpose the C++ interface provides the two preprocessor macros INSTANCE(C,I) and EXPORT(C,I). Application of the first macro creates an component instance of class C with the name I. This name can then be used to associate this component instance with a foreign architecture as described in section 3.3. Application of the second macro creates a set of C wrapper functions that encapsulate instance construction and all instance method calls from the simulation engine side.

4.   EXAMPLE

After discussing the system architecture in the previous chapter we now present a complete example of a mixed-language VHDL/JAVA design. Due to the limited space in this publication the design is rather small. The example is a modulo-16-counter that is triggered by the positive edges of a clock signal generated by a clock generator. Figure 6 shows the JAVA implementation of the counter component and the VHDL implementations of the top-level model structure, the clock generator, and the counter wrapper.

The JAVA counter component is bound to the corresponding VHDL architecture BEH by the FOREIGN attribute specification in lines 36 to 38. This binding defines the component interface which is determined by the port definitions of entity COUNTER in lines 30 to 32. The JAVA component declares matching ports in lines 45 to 47 that are mapped to the entity ports in method init (lines 52 to 54). Invalid mappings of port signals result in an exception and generate a failure message which stops simulation (line 58). During simulation the counter will be reset by every positive edge occurring on the reset signal as shown in lines 64 to 67. This signal has a higher precedence than signal clock thus modelling an asynchronous reset

```
01  entity TOP is                         41  import vhdl.java.comp.*;
02  end TOP;                              42  import vhdl.java.typ_std.*;
03                                        43
04  architecture STRUC of TOP is          44  public class VJCounter extends VJArchitecture {
05      signal reset : BIT := '0';        45      private VJInterfaceSignalIn   _clock;
06      signal clock : BIT := '0';        46      private VJInterfaceSignalIn   _reset;
07      signal count : INTEGER := 0;      47      private VJInterfaceSignalOut  _out;
08  begin                                 48      private long                  _value;
09      CLK: entity WORK.CLOCK(BEH)       49
10          port map( clock );            50      public void init( VJInterfaceObject objs[]) {
11      CNT: entity WORK.COUNTER(BEH)     51          try {
12          port map( clock, reset, count);  52          _clock= registerSignalIn( objs[ 0]);
13  end STRUC;                            53          _reset= registerSignalIn( objs[ 1]);
14                                        54          _out=   registerSignalOut( objs[ 2]);
15  entity CLOCK is                       55          _value= 0;
16      port( clock : out BIT);           56          }
17  end CLOCK;                            57          catch( VJEXception vjExn) {
18                                        58          writeFailure( vjExn.getMessage());
19  architecture BEH of CLOCK is          59          }
20  begin                                 60      }
21      process begin                     61
22          wait for 50 ns;               62      public void run( long  time) {
23          clock <= '1';                 63          try {
24          wait for 50 ns;               64          if( _reset.event()
25          clock <= '0';                 65          && _reset.hasValue( VJTypeBit.val( '1'))){
26      end process;                      66              _value= 0;
27  end BEH;                              67          }
28                                        68          else {
29  entity COUNTER is                     69          if( _clock.event()
30      port( clock : in BIT;             70          && _clock.hasValue( VJTypeBit.val('1'))){
31            reset : in BIT;             71              _value= (_value + 1) % 16;
32            count : out INTEGER);       72              _out.write( timc,
33  end COUNTER;                          73                     VJTypeInteger.val( _value));
34                                        74          }
35  architecture BEH of COUNTER is        75          }
36      attribute FOREIGN of              76          }
37      BEH : architecture is             77          catch( VJEXception vjExn) {
38      "JAVA:VJCounter";                 78          writeFailure( vjExn.getMessage());
39  begin                                 79          }
40  end BEH;                              80      }
                                          81  }
```

Figure 6. Example source code

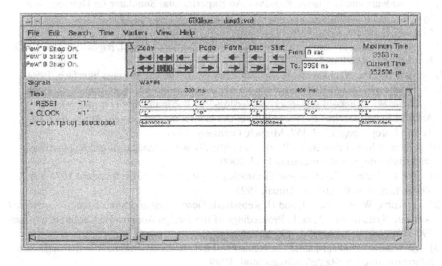

Figure 7. Example simulation results

mechanism. The counter is incremented by every positive edge occurring on the clock signal and the resulting new value is written to the out signal using method write (lines 69 to 74). Thereby, the time stamp of the generated event is set to the current time as specified by parameter time of method run. The resulting output waveform is shown in figure 7.

5.   SUMMARY

We presented a new mixed-language simulation environment for the languages VHDL, JAVA, and C++. The environment is based upon an existing VHDL simulator that was extended by open object-oriented JAVA and C++ interfaces. Obviously, this system architecture lends itself to a VHDL-centric modelling style. However, the advantage of this approach is the resulting clear simulation semantics that is solely based on the semantic principles of a single language, i.e. VHDL.

BIBLIOGRAPHY

[1]   The SLDL Home Page. http://www.inmet.com/SLDL, 2000.
[2]   A. Jerraya. Multilanguage Specification for System Design. In Proceedings of the SLDL Workshop, Grenoble, 1997.
[3]   A. Jerraya and R. Ernst. Multi-language System Design. In Proceedings of the Conference on Design, Automation and Test in Europe, pages 696-699, Munich, Germany, 1999.
[4]   E. Lee and A. Sangiovanni-Vincentelli. Comparing Models of Computation. In Proceedings of the International Conference on Computer Aided Design, San Jose, 1996.
[5]   P. Schwarz and C. Clauss. KOSIM – An Experimental Simulator for Heterogeneous Systems. In Proceedings of the Workshop on Mixed-Signal and Mixed-Level Simulation, Munich, Germany, 1999.
[6]   Mentor Graphics Corporation. ModelSim EE/SE Users Manual, pages 297-334. VHDL Foreign Language Interface, 1999.
[7]   Synopsys Inc. VSS Interfaces Manual, Chapter 3. The C-Language Interface, 1999.
[8]   T. Johnson. Mixed-HDL Simulation: How does it work. VHDL-Times, 8(2), 1999.
[9]   H. Sasaki. A Formal semantics for Verilog-VHDL Simulation Interoperability by Abstract State Machine. In Proceedings of the Conference on Design, Automation and Test in Europe, pages 353-357, Munich, Germany, 1999.
[10]   Cadence Signal Processing Workbench. http://www.cadence.com/technology/hwsw/ products/cierto_signal_processing.html, 2000.
[11]   The Institute of Electrical and Electronics Engineers Inc. IEEE Standard 1076-1993. VHDL Language Reference Manual, 1993.
[12]   T. Kuhn, W. Rosenstiehl, and U. Kebschull. Description and Simulation of Hardware / Software Systems with Java. In Proceedings of the Design Automation Conference, pages 790-793, New Orleans, USA, 1999.
[13]   A. Windisch and J. Mades. The Infineon VHDL-AMS Environment. http://herkules. informatik.tu-chemnitz.de/ive/index.html, 1999.
[14]   T. Schneider and J. Mades. The Infineon VHDL-AMS Environment. http://www. microelectronic.e-technic.tu-darmstadt.de/research/CAD/IVE/ive.html, 1999.
[15]   A. Windisch, W. Ecker, C. Hammer, T. Schneider, J. Mades, and K. Yang. An Adaptable VHDL-AMS Compiler Front-End. In Proceedings of the Forum on Design Languages, pages 38-44. IEEE Computer Society, IEEE Press, Los Alamitos, 1999.
[16]   J. Mades, T. Schneider, A. Windisch, and W. Ecker. Elaboration of Hierarchical VHDL-AMS Models for Mixed-Signal Simulation. In Proceedings of the International HDL Conference and Exhibition, pages 111-117, San Jose, USA, 2000.

Analogue circuit synthesis from VHDL-AMS

Tom J Kazmierski and Fazrena A Hamid

Department of Electronics and Computer Science, University of Southampton, United Kingdom

Abstract: The paper presents the evolution of analogue synthesis techniques and present recent research results into automated analogue synthesis in the context of VHDL-AMS. The emergence of VHDL-AMS provides a basis for a new approach to analogue and mixed-signal circuit synthesis. Like digital VHDL, VHDL-AMS supports process-level parallelism and provides constructs to describe process communication and signal assignments. This gives rise to a development of architectural analogue synthesis techniques that would be analogous to the well-established methods in digital synthesis based on high-level VHDL or Verilog descriptions. Two sample synthesis systems based on VHDL-AMS, i.e. VASE and NEUSYS are presented.

1. INTRODUCTION

Integrated circuit synthesis is defined as the process of automatic generation of integrated circuit layout masks from a high-level description of circuit behaviour. In the field of architectural synthesis of digital circuits

47

P.J. Ashenden et al. (eds.), System-on-Chip Methodologies & Design Languages, 47–56.
© 2001 *Kluwer Academic Publishers. Printed in the Netherlands.*

behavioural specifications are usually written in a high-level hardware description language, such as Verilog or VHDL. The classical two-stage synthesis process comprises an automatic translation of the behavioural description to a circuit structure, which is then followed by silicon compilation to produce fabrication masks or field programmable gate arrays. The process of generating silicon layout from high-level descriptions is analogous to translating a computer program written in a high-level programming language to a machine code.

The advances in integrated circuit technology have led to the growing popularity and a decrease in cost of mixed-signal ASICs (Application Specific Integrated Circuits), which comprise both analogue and digital circuit blocks. Significant application areas of mixed-signal ASICs include for example signal processing, mobile telephony and computer networking. However, the development of CAD tools in recent years has concentrated mainly on automatic design of digital circuits while the automation of analogue designs remains largely a heuristic and a labour-intensive task. There are two main reasons for the significant popularity high-level digital synthesis has gained during the last decade. Firstly, high-level digital design specifications are now very well supported by suitable hardware description languages, which allow design constraints and functional descriptions to be clearly stated. Secondly, methods and algorithms for digital synthesis and design optimisation have been well established and are supported by a large number of software tools used on an industrial scale throughout the integrated circuit design community. The development of appropriate synthesis methodologies to support mixed-signal ASIC designs is still lagging. While digital designs are now fully automated and can be delivered extremely quickly, the analogue part of a typical ASIC still needs to be designed manually. With the advent of digital synthesis tools and semi-custom layout design techniques, analogue ASIC blocks may now consume 90% of the overall design time, while using only 10% of the silicon area [1]. One of the obvious difficulties has been the absence of a mixed-signal high-level hardware description language commonly accepted as a standard throughout the CAD industry.

The IEEE DASC (Design Automation Standards Committee) has recently formally adopted the 1076.1 standard for VHDL informally called VHDL-AMS (VHDL hardware description language with analogue and mixed-signal extensions) [2]. VHDL-AMS extends the modelling power of VHDL to the domain of continuous and discrete-continuous systems. The new hardware description standard is expected to advance behavioural modelling of very large and complex systems such as, for example, fuzzy control systems or artificial analogue VLSI neural networks. Analogue SPICE-like simulators, which have so far been used for most VLSI analogue simulations, do not support behavioural modelling and do not lend themselves easily to system-level modelling. Following the formal release of the new standard by the IEEE, commercial simulators are already available from Mentor Graphics, Analogy and FTL.

2.   STATE OF THE ART IN ANALOGUE SYNTHESIS

The evolution of analogue circuit synthesis started with attempts to base the synthesis process on equations and mathematical optimisations. Those early efforts failed in practical implementations due to the size and complexity of the equations driving the underlying synthesis tools [1]; in addition, these methods are limited in accuracy and difficult to automate.

Subsequent to the early equation-based synthesis tools, is the method in which the equations are formulated automatically using design databases obtained from human experts. IDAC (Interactive Design for Analog Circuits) [3] is one the earliest knowledge-based analogue circuit synthesis tools. As an input to the system, the user has to specify building block parameters and choose a topology from the library. The library contains schematics of verified analogue circuits and their formal description. The schematics consist of groups of blocks such as op amps, comparators and oscillators. One of the major drawbacks of this type of synthesis is the amount of time and effort invested in extracting and compiling the human knowledge from analogue designers' expertise into a database suitable for the use of the tools. Also, simulations need to be carried out a number of times to correctly evaluate the design performance.

Another successful knowledge-based synthesis tool, OPASYN [4] was developed specifically for synthesis of operational amplifiers. It uses an optimisation algorithm, which is based on algebraic design equations taken from the domain-specific design knowledge. The user provides op amp performance specifications, which are taken by the tool to select the most suitable op amp that it has in its internal database. This database contains design knowledge necessary to make the correct selection of op amp circuit.

An approach to this is to include simulation into the equation-based system, as used in OAC [5]. The simulator is used as an optimiser, which is a post-processor. However, there are problems associated with this, as the optimisation depending on the simulation creates uncertainty in the selection of starting point of synthesis, also, the amount of time involved in running the simulation.

There are also synthesis tools that are not equation-based or simulation-based. For example, ASTRX/OBLX [1], uses simulated annealing for optimisation. Another synthesis tool, BLADES [6] is a knowledge-based system, which uses an expert database. It applies artificial intelligence and reasoning, which combine the application of formal and intuitive design knowledge. It uses an expert system that stores two types of knowledge – formal, based on mathematical techniques and heuristic, intuitive knowledge based on experience.

STAIC [7] is an interactive design tool that synthesises CMOS and BiCMOS analogue integrated circuits. It features the use of a description language for entering hierarchical circuit descriptions and a solver unit which dynamically integrates analytical model equations across the levels of hierarchy. The synthesis method is the successive solution refinement. The analogue hardware description language used in STAIC classifies circuits as being categorist, specific, device or layout. This classification is based on the circuit complexity level.

Several tools have been developed for the synthesis at system-level of mixed, analogue and digital circuits. ARCHGEN [8] deals specifically with filter systems; KANDIS [9] is a general tool for mixed-signal synthesis, whereas VASE [10] is a synthesis environment for mixed-signal circuits from VHDL-AMS. VASE is discussed in more detail in the next section.

Traditionally, synthesis of analogue circuits relies on systems that accept input in terms of performance specifications and process parameters. Only very recently researchers have developed tools that accept analogue descriptions in a high-level hardware description language as its input, as in the case of KANDIS ,VASE and NEUSYS [11,12].

Table 1. Summary of Analogue Circuit Synthesis Systems

	Name	Type	Origin and Description	Year
1	IDAC	KB	Swiss Center for Electronics and Microtechniques, Switzerland. Users select a topology from a library. 1991's: open tool for design reuse.	1987, 1991
2	AN-COM	KB	General Electric Company, New York, USA. Domain knowledge is used for successive decomposition of circuit specification.	1988
3	CAMP	KB	Univ. of Southern California, USA. Uses iterative self-reconstructing technique and circuit simulation for a flexible architecture.	1988
4	DELIGHT.SPICE	OB	Harris Corporation, USA. Utilises a SPICE simulator as the optimisation core [13].	1988
5	OASYS	KB	Carnegie Mellon Univ., USA. Top-down hierarchical structure in knowledge application [14,15]	1989
6	BLADES	KB	AT&T Bell Labs., USA. Uses artificial intelligence to combine formal and intuitive knowledge.	1989

	Name	Type	Origin and Description	Year
7	OPASYN	KB	Univ. of California, Berkeley, USA. Silicon compilation of op amps.	1990
8	ASAIC	OB	Katholieke Universiteit Leuven, Belgium. Features a symbolic analysis programme, ISAAC and an optimiser, OPTIMAN [16,17,18].	1990
9	CHIPAIDE	KB	Imperial College, UK. Uses a hierarchical approach to produce first-cut circuit topology [19].	1990
10	OAC	OB	Kyoto Univ., Japan. CMOS op amp compiler which runs a simulation-based optimiser as a post-processor.	1990
11	STAIC	OB	Univ. of Waterloo, Ont., Canada. Uses description language in its multilevel modelling scheme. Synthesis uses successive solution refinement technique.	1992
12	MINLP-Maulik	OB	Carnegie Mellon Univ., USA. Allows simultaneous circuit topology and parameter selection [20].	1992
13	ARCHGEN	MS	Vanderbilt Univ., USA. Synthesis of filter systems from behavioural specifications.	1995
14	KANDIS	KB (MS)	Johann Wolfgang Goethe Univ., Frankfurt. Translates hybrid-VHDL into intermediate representation (KIR graph), which are then used by a high-level synthesis tool and an estimator. An expert system plays a vital role in the synthesis of the analogue part.	1995
15	ASTRX/OBLX	OB	Carnegie Mellon Univ., USA. Uses asymptotic waveform evaluation (AWE) to evaluate circuit performance and simulated annealing for optimisation.	1996
16	VASE	OB (MS)	Univ. of Cincinnati, USA. Behavioural VHDL-AMS specifications are compiled to obtain a hierarchical intermediate representation. An architecture generator and performance estimator is part of the synthesis environment.	1997
17	NEUSYS	OB	Univ. of Southampton, UK. An architecture generator translates the parse trees obtained from behavioural VHDL-AMS specifications into intermediate netlists for subsequent optimisation.	2000

KB: Knowledge based, OB: Optimisation based, MS: Mixed-signal system

3.   LOOKING INTO THE FUTURE

The emergence of VHDL-AMS provides a basis for a new approach to analogue synthesis. Topology compilers can be developed to translate high-level behavioural equations from VHDL-AMS to structural netlist of analogue cells in a way that is analogous to digital circuit synthesis from VHDL or Verilog. VASE (VHDL-AMS Synthesis Environment) [10] is the first analogue synthesis system that uses VHDL-AMS behavioural descriptions. The synthesis output is a sized circuit-level net list, which satisfies the performance constraints and minimises the ASIC area. This is achieved by using a two-layered design space exploration [21]. In the exploration phase, there are two interacting steps, i.e. architecture generation, constraint transformation and component synthesis. The latter step interacts with the analogue performance estimator (APE) in the estimation phase.

The input to VASE is an intermediate VHIF representation, which is obtained by compiling the VHDL-AMS specification of a system [22]. In VHIF, continuous-time behaviour is represented as signal-flow graphs. The architecture generator gets the topologies from a library. This analogue components library contains unsized topologies of commonly used analogue circuits. APE accepts design parameters and the corresponding topology of an analogue circuit and determines the performance parameters and the sizes of circuit elements.

Another VHDL-AMS based synthesis system, named NEUSYS, has been developed to aid the design of large artificial neural networks (ANNs). Silicon implementations of ANNs, especially large analogue networks of the type required in fuzzy logic and control systems, might provide an important area for analogue synthesis applications. Neural networks are a particularly interesting field of study because of their inherent hierarchical structure and the behavioural description of individual neurons. Biologically inspired ANNs are often used in robotic arm control systems as they reflect the basic structure and behaviour of the human spinal cord [12], focusing on the motor control of the superior limbs. NEUSYS is an automated synthesis system that converts VHDL-AMS descriptions of highly interconnected networks of neural cells into HSPICE net lists. The main motivation for automated synthesis of large analogue neural networks is to overcome the notorious difficulties and prohibitive simulation times that occur during post-layout verification of very large analogue integrated circuits.

A very simple example of an ANN structure is presented in figure 2 in which a group of antagonist muscles are controlled by a neural network, which consists of a pair of neurons called motor-neurons (M1 and M2) and another pair called alfa inter neurons (I1 and I2). This simple network is controlled by a double input (A1, A2).

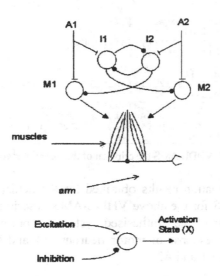

Figure 1. A model of a neural system controlling a pair of antagonist muscles.

A suitable VHDL-AMS description for the neural system is shown in **Figure 2 and Figure 3.**

```
-- Definition for Motor Neuron model
entity Motor_neuron is
    port (quantity M: out real; A: in real; IaI_opp: in real);
end entity Motor_neuron;

architecture simple_1 of Motor_neuron is
    quantity lambda: real:= 3.0;
begin
-- initial conditions
break M => 0.0;
-- behaviour equation:
M'dot == (lambda-M)*A - (M+1.6)*(0.2+IaI_opp);
end architecture simple_1;

-- Definition for Alfa Inter Neuron model
entity Inter_neuron_alfa is
    port (quantity IaI: out real;  A: in real; IaI_opp: in real);
end entity Inter_neuron_alfa;

architecture simple_1 of Inter_neuron_alfa is
begin
    -- initial conditions
    break IaI => 2.0;
    -- behaviour equation:
    IaI'dot == (10.0-IaI)*A - (IaI+1.0)*(1.0+IaI_opp);
end architecture simple_1;
```

Figure 2. VHDL-AMS description of the neurons.

```
-- Structural Design Entity
entity network_1 is
    port (A1: in real; A2: in real; out1: out real; out2: out real);
end entity network_1;

-- Architecture
architecture behaviour of network_1 is
    -- use work.neurons_1.all;
    quantity IaI11: real;  quantity IaI12: real;
    quantity IaI21: real;  quantity IaI22: real;
begin
  C_IaI1: Inter_Neuron_alfa
    port map (A=>A1, IaI=>IaI12, IaI=>IaI11, IaI_opp=>IaI21);
  C_IaI2: Inter_Neuron_alfa
    port map (A=>A2, IaI=>IaI21, IaI=>IaI22, IaI_opp=>IaI12);
  C_M1: Motor_Neuron   port map (A=>A1, IaI_opp=>IaI22, M=>out1);
  C_M2: Motor_Neuron   port map (A=>A2, IaI_opp=>IaI11, M=>out2);
end architecture behaviour;
```

Figure 3. VHDL-AMS description of the neural network.

The HSPICE simulation results obtained for the netlist, which has been generated by NEUSYS for the above VHDL-AMS description [11], showing the dynamic behaviour of the synthesised system are presented in Figure 4. The graphs show the response of motor neurons M1 and M2 for step-wise changes of excitations A1 and A2.

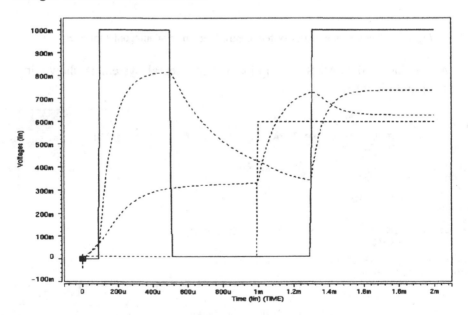

Figure 4. HSPICE simulation results of the muscle control neural network.

4. CONCLUSION

Methods and algorithms for digital synthesis and design optimisation have been well established and are supported by a large number of software tools used on an industrial scale throughout the integrated circuit design community. While digital designs are now fully automated, the development of appropriate synthesis methodologies to support mixed-signal ASIC designs is still lagging.

The main advantage of the new VHDL-AMS standard in the contexty of analogue synthesis is that topology compilers can now be developed to translate high-level behavioural equations from an HDL to structural netlist of analogue cells. VHDL-AMS based synthesis is likely to provide, in many cases, an elimination of hand-crafted topologies required by most existing analogue synthesis tools. One can predict that a combination of heuristic and systematic algorithms will need to be developed in order to optimise synthesised netlists for both circuit size and functional accuracy.

REFERENCES

[1] E.S. Ochotta, R.A. Rutenbar, L.R. Carley, "Synthesis of High-Performance Analog Circuits in ASTRX/OBLX", *IEEE Trans. CAD*, vol. 15, pp. 273-294, Mar. 1996.

[2] Design Automation Standards Committee of the IEEE Computer Society, "IEEE Standard VHDL Language Reference Manual (Integrated with VHDL-AMS changes)", IEEE Std 1076.1. July, 1999.

[3] M.G.R. Degrauwe, O. Nys, J. Dijkstra, S. Bitz, B.L.A.G Goffart, E.A. Vittoz, S. Cserveny, C. Meixenberger, G.V.D.S. Stappen, H.J. Oguey, "IDAC: An Interactive Design Tool for Analog Circuts", *IEEE Journal of Solid-State Circuits*, vol. 22, no. 6, pp. 1106-1115, 1987.

[4] H.Y. Koh, C.H. Sequin, P.R. Gray, "OPASYN: A Compiler for CMOS Operational Amplifiers",.*IEEE Transactions on Computer Aided Design of Integrated Circuits and Systems*, vol. 9, no. 2, pp. 113-125, 1990.

[5] H. Onodera, H. Kanbara, K. Tamaru, "Operational-amplifier compilation with performance optimization", *IEEE J. Solid-State Circuits*, vol. 25, pp 466-473, Apr., 1990.

[6] F. El-Turky, E.E. Perry, "BLADES: An Artificial Intelligence Approach to Analog Circuit Design", *IEEE Transactions on Computer Aided Design of Integrated Circuits and Systems*, vol. 8, no. 6, pp. 680-692, 1989.

[7] J.P. Harvey, M.I. Elmasry, B. Leung, "STAIC: An interactive framework for synthesizing CMOS and BiCMOS analog circuits", *IEEE Transactions on Computer-Aided Design*, vol. 12, pp. 1402-1418, Nov. 1992.

[8] B. Antao, A. Brodersen, "ARCHGEN: Automated Synthesis of Analog Systems", *IEEE Transactions on VLSI*, vol. 3, no. 2, pp. 231-244, June 1995.

[9] P. Oehler, C. Grimm, K.Waldschmidt, "KANDIS – A Tool for Construction of Mixed Analog/Digital Systems", *European Design Automation Conference*, (Brighton, UK), Sept. 1995.

56

[10] R. Vemuri, A. Nunez-Aldana, N. Dhanwada, A. Doboli, P. Campisi, S. Ganesan, "Analog System Performance Estimation in the VASE", *Proc. EETimes Analog And Mixed-Signal Applications Conference*, pp. 65-70, July 1998.

[11] G. Domenech-Asensi , T. J. Kazmierski, "Automated synthesis of high-level VHDL-AMS analog descriptions", *First On Line Symposium For Electronic Engineers*, Sept. 2000, http://techonline.com/osee/.

[12] G. Doménech-Asensi, R. Ruiz-Merino, T.J. Kazmierski, "Automatic synthesis of analog systems using a VHDL-AMS to HSPICE translator", *DCIS'2000*, Montpelier, Nov. 21-24 2000.

[13] W. Nye, D. Riley, A. Sangiovanni-Vincentelli, A. Tits, "DELIGHT.SPICE: an optimization-based system for the design of integrated circuits", *IEEE Transaction on Computer-Aided Design*, vol. 7, no. 4, pp. 501-519, April 1988.

[14] R. Harjani, R.Rutenbar, L. Carley, "OASYS: a framework for analog circuit synthesis", *IEEE Transaction on Computer-Aided Design*, vol. 8, no. 12, pp. 1247-1266, December 1989.

[15] E. Ochotta, "The OASYS virtual machine: Formalizing the OASYS analog synthesis framework", Master's thesis, Carnegie Mellon Univ., 1989.

[16] G. Gielen, H. Walscharts, W. Sansen, "Analog Circuit Design Optimization Based on Symbolic Simulation and Simulated Annealing", *IEEE Transaction on Solid-State Circuits*, vol. 25, no. 3, pp. 707-713, June 1990.

[17] G. Gielen, K. Swings, W. Sansen, "An Intelligent Design System for Analogue Integrated Circuits", *proc. European Design Automation Conference*, pp. 169-173, 1990.

[18] K. Swings, G. Gielen, W. Sansen, "An intelligent analog IC design system based on manipulation of design equations", *proc. CICC*, pp. 8.6.1-8.6.4, 1990.

[19] C.A. Makris, C.M. Berrah, X. Xiao, M. Singha, A.A. Ilumoka, J. Stone, C. Toumazou, P.Y.K. Cheung, R. Spence, "CHIPAIDE: A New Approach to Analogue Integrated Circuit Design", *IEE Colloqium on 'Analogue VLSI'*, (Digest No.073), pp. 1/1-1/11, 1990.

[20] P.C. Maulik, L.R. Carley, R.A. Rutenbar, "A mixed-integer nonlinear programming approach to analog circuit synthesis", *Proc. Design Automation Conf.*, pp. 698-703, June 1992.

[21] A. Doboli, N. Dhanwada, A. Nunez-Aldana, S. Ganesan, R. Vemuri. "Behavioral Synthesis of Analog Systems using Two-Layered Design Space Exploration", *Proceedings of the 36th Design Automation Conference*, June 1999.

[22] A. Doboli, R. Vemuri, "The Definition of a VHDL-AMS Subset for Behavioral Synthesis of Analog Systems", *1998 IEEE/VIUF International Workshop on Behavioral Modeling and Simulation (BMAS'98)*, Oct. 1998.

FORMAL VERIFICATION

SYMBOLIC SIMULATION AND VERIFICATION OF VHDL WITH ACL2

Dominique Borrione, Philippe Georgelin
Laboratoire TIMA-UJF, 46, Avenue Felix Viallet, 38031 Grenoble Cedex
{Dominique.Borrione, Philippe.Georgelin}@imag.fr

Vanderlei Moraes Rodrigues
Instituto de Informática, UFRGS at Porto Alegre, Brazil
vandi@inf.ufrgs.br

Abstract *ACL2 is a theorem prover to reason about specifications written in a quantifier-free, first-order logic. The input langage is Common Lisp. We defined the semantics of a synthesizable VHDL subset in the logic, and describe an automatic method (and the corresponding tool) to build an ACL2 model of an abstract behavioral VHDL description. This model is executable and efficient, so the designer can simulate it on the same tests as with a conventional simulator, and get confidence in it. Using the theorem prover, we may prove properties that otherwise would require a large or infinite number of simulation runs. Finally, we may also perform symbolic simulation, i.e. we can execute the design on symbolic input data, which is another way to cope with excessively large or infinite sets. Symbolic simulation is an automatic process, so it is easier than theorem proving. We stress the fact that all three tasks are performed on the same representation of a VHDL design.*

Keywords: Theorem prover, VHDL, ACL2, Symbolic simulation

Introduction

ACL2 [8] is the descendant of the highly successful NQTHM [3] theorem prover of Boyer and Moore. ACL2 reasons on a quantifier-free first order logic, written in an applicative subset of Common Lisp. In practice, it is a tool to state and check properties of Lisp programs that

P.J. Ashenden et al. (eds.), System-on-Chip Methodologies & Design Languages, 59–69.

usually model some computational system. Compared to other logics such as HOL [6] or PVS [12], Lisp is a rather restricted specification language. However, it allows for some highly powerful and automated proof techniques (e.g. induction) which are not fully automated in other tools, relieving the ACL2 user from providing the proof engine with the detailed inference steps. Furthermore, since Lisp is a compiled programming language, all ACL2 specifications are executable in an efficient way, and the ACL2 models are reasonably easy to build and understand.

We thus explore the verification of VHDL designs using ACL2. We are interested in high-level, behavioral descriptions on which other verification techniques such as finite model-checking [9] fail. We build a translator that takes a VHDL design and generates the corresponding ACL2 model. It comprises a set of functions and theorems describing the simulation behavior of the VHDL design. This model can be used for three tasks.

1 Since ACL2 functions are in ordinary Lisp, and modern Lisp compilers produce good code, the model can be efficiently *executed* on test cases, reproducing the effect of a standard VHDL simulator. Actually, the designer is expected to check that the model *execution* and the VHDL *simulation* provide identical results on the same test cases. This is important, because it gives confidence to the engineer that the ACL2 model is faithful to the original VHDL.

2 The ACL2 model can be used to *formally verify* properties of the VHDL design. The desired properties are stated as theorems on the design representation, and proved using the theorem prover engine with the aid of lemmas generated with the ACL2 model. Due to the mathematical nature of the proof process, the proof of a single property may describe a very large or even infinite number of cases, which cannot be done using ordinary simulation.

3 ACL2 may also perform *symbolic simulation*, which corresponds to simulating the VHDL design on input mathematical variables $(x, y, \ldots)$, representing arbitrary values in some domains. Thus, a single simulation run may also deal with a very large or infinite number of cases. This method is more restricted than theorem proving, but it is easier and fully automated. It is implemented by a set of theorems (also generated by the translator from VHDL to ACL2) that "direct" the theorem prover engine to symbolically simulate the design, taking advantage of the advanced techniques built in ACL2 to compare and simplify symbolic expressions [10].

```
ENTITY mysystem IS
   PORT (arg: IN natural;
         start: IN bit;
         res: OUT natural;
         done: OUT bit;
         clk: IN bit);
END mysystem;
ARCHITECTURE arc OF mysystem IS
   SIGNAL op1, op2, resmult: natural;
   SIGNAL reqmult, ackmult: bit;
BEGIN
   mult: PROCESS
   BEGIN
      WAIT UNTIL clk = '1';
      IF reqmult = '1' THEN
         resmult <= op1 * op2;
      END IF;
      ackmult <= reqmult;
   END PROCESS mult;
   control: PROCESS
      ...
   END PROCESS control;
END arc;
```

```
(_entity mysystem
 :port (arg :in natural
        start :in bit
        res :out natural
        done :out bit
        clk :in bit))
(_architecture arc :of mysystem :is
  ((_signal op1 :type natural)
   (_signal op2 :type natural)
   (_signal resmult :type natural)
   (_signal reqmult :type bit)
   (_signal ackmult :type bit))
 :begin
 ((_process mult :is () :begin
    ((_wait :until (equal clk 1))
     (_if (equal reqmult 1) :then
        ((_<= resmult (* op1 op2))))
     (_<= ackmult reqmult)))
  (_process control ...)))
```

Figure 1.1.   A VHDL design and its representation in the intermediate format

1.   ACL2 MODELS OF VHDL DESIGNS

VHDL is large and complex language, so we concentrate on a behavioral synthesis subset of VHDL currently being standardized [7]. This subset excludes physical time and non-discrete types. We further require that processes be synchronized on a single clock edge, and that they be put in a normal form with a single, starting wait statement. Under these conditions, we may identify the simulation step with the clock cycle. Presently, a design is represented by an entity followed by its associated architecture composed of several processes, as exemplified in the left column of Figure 1.1.

We employ a prefix, fully-parenthesized, Lisp-like intermediate format to represent VHDL designs. This representation closely mirrors the original design, yet it is more amenable to treatment by ACL2. Standard VHDL source code is translated into this format by a C program. From this format, the semantic model of the design is produced automatically.

We wrote a set of macros and functions so that VHDL designs in the intermediate format can be directly read into ACL2 as an ordinary program, and the standard Lisp macro expander translates these designs into the corresponding functions and theorems. Since the intermediate format is syntactically very close to the original program, it gives further confidence to the engineer that the design representation for proofs is faithful to standard VHDL (see Figure 1.1 right column). In this format, we need to slightly adapt the VHDL syntax and keywords to match Lisp conventions, and to avoid name clashes with predefined macros.

To build an ACL2 representation of a design, we cannot perform a straightforward translation from VHDL to Lisp, because all ACL2 functions are required to be total (terminating), while the VHDL simulation cycle may end abruptly (e.g. on an `assert...severity error`). Following the work of Moore [10], we represent each architecture in a purely applicative setting as an abstract machine mapping states to states.

A *state* is a list of values, representing a snapshot of the entity-architecture declared objects: variables, local signals, and interface signals. More precisely:

- For each `variable` declared in a process, there is a corresponding element in the state list, that holds the variable value.

- For each `signal`, there is one state element storing the signal *current* value, and another state element holding the signal *next* value. There is a single next value, which corresponds to the signal driver, due to the restrictions imposed on the time model.

For instance, using the element name to represent its value, the state of the design in Figure 1.1 is the list (arg start ... arg+ start+ ...). By convention, the next value of signal x is named $x+$.

In order to ease state manipulation, each `variable` or `signal` declaration in a VHDL design generates several functions in the ACL2 model. For instance, for each element *elem*, function get-*elem* fetches the value of *elem*, and set-*elem* updates it, thus producing a new state. To avoid ambiguities, local names are prefixed with architecture and process names, when appropriate.

As an example, in the architecture `arc` of Figure 1.1, assume the current value of signal `resmult` holds position 8, and its next value holds position 15 in the model state `st`. The following state manipulation functions are automatically produced with the model:

;; access current value:	;; assign next value:
(defun get-arc.resmult (st)	(defun set-arc.resmult+ (x st)
(nth 8 st))	(set-nth 15 x st))

The function (nth n l) returns the element of list l in position n, and (set-nth n x l) generates a new list identical to l, except that position n holds x.

The ACL2 model uses the notion of state explicitly to describe the effect of statements. The execution of a VHDL statement in a given state is represented by a function that computes the corresponding result state. For instance, the execution of the variable assignment $x := e$ in a state st is translated to (set-x e' st), where e' corresponds to expression e where references to variables and signals are replaced by accesses to

```
(defun arc.mult-cycle (st0)                    ;; body of process "mult"
  (let* ((st1 (if (equal (get-arc.reqmult st0) 1)
                  (set-arc.res+ (* (get-arc.op1 st0) (get-arc.op2 st0)) st0)
                  st0))
         (st2 (set-arc.ackmult+ (get-arc.resmult st1) st1)))
    st2))
(defun arc.control-cycle (st0)                 ;; body of process "control"
  ...)
(defun arc-update-signal (st)                  ;; update signals
  (put-arc.arg (get-arc.arg+ st)
    (put-arc.start (get-arc.startx+ st)
      ... st)))
(defun arc-cycle (st)                          ;; one simulation step
  (arc-update-signal (arc.mult-cycle (arc.control-cycle st))))
(defun arc-simul (n st)                        ;; several simulation steps
  (if (and (intergerp n) (> n 0))
      (arc-simul (- n 1) (arc-cycle st))
      st))
```

Figure 1.2.   Fragment of the ACL2 model for the VHDL design of Figure 1.1

the corresponding state elements in *st*. Likewise, a signal assignment *x* <= *e* becomes an update (set-*x*+ *e′ st*) to the next signal value.

A sequence of VHDL statements is tranlsated as the composition of the corresponding functions using the let* operator, and conditional statements have a straightforward representation. Loop statements require the introduction of intermediate functions to perform the iteration. Operations on more complex data types, such as vectors and vector slices, also require the introduction of new functions. More details on the generation of user-types and arrays can be found in [1].

For each process of the VHDL design, the ACL2 model includes a function that describes the new state produced by one execution step of this process. This function body results from a syntax-directed translation of the statements in the process body according to the above method. For instance, in Figure 1.2, functions arc.mult-cycle (st0) and arc.control-cycle (st0) model the effect of one execution cycle of processes mult and control on any current state st0 of the architecture arc. Notice that the process functions only update local variables and next signal values.

To model the simulation cycle of an architecture arc (see Figure 1.2), we define a function arc-cycle that calls the functions corresponding to each process and then calls function arc-update-signal to replace the current value of each signal with its (possibly resolved) next value. Finally, the transition function representing the architecture is a recursive function arc-simul that repeats the architecture simulation cycle *n* times, returning the corresponding state. The number *n* is a decreasing argument, ensuring that the computation terminates.

Using the above functions of the ACL2 model, we may simulate the VHDL design. Let s be a state with the appropriate input values. The evaluation of (arc-simul n s) returns the state after n simulation steps. But this ACL2 model can also be used to formally verify properties of the design, and to perform symbolic simulation.

2. FORMAL VERIFICATION AND SYMBOLIC SIMULATION

```
ARCHITECTURE fact OF mysystem IS
SIGNAL ct, state: natural := 0;
BEGIN|
   doit: PROCESS
    BEGIN|
       WAIT UNTIL clk = '1';
       CASE state IS
          WHEN 0 => IF start = '1' THEN
                       res <= 1; ct <= arg; state <= 1; END IF;
          WHEN 1 => IF ct > 0 THEN
                       res <= res * ct; ct <= ct - 1;
                    ELSE
                       done <= '1'; state <= 2; END IF;
          WHEN 2 => IF start = '0' THEN
                       done <= '0'; state <= 0; END IF|
          WHEN OTHERS => state <= 0;
       END CASE;
    END PROCESS;
END fact;                Figure 1.3. A simple design
```

In the following, we consider a simple reactive system (Figure 1.3). In this design, signal start activates a state machine that stores in res the factorial of arg. When this computation is finished, it activates done. As discussed earlier, the ACL2 model for this design includes the following functions to describe its simulation behavior:

```
(defun fact-cycle (st)          (defun fact-simul (n st)
  (fact-update-signal             (if (and (intergerp n) (> n 0))
    (fact.doit-cycle st))))          (fact-simul (- n 1) (fact-cycle st))
                                      st))
```

Using the ACL2 model of a design, we may formally state and prove the properties of the design. For instance, the theorem below indicates that the design above is correct:

```
(defthm fact-correct
  (implies (and (naturalp x) (equal st (list x 1 0 0 1 0 0 ...)))
           (equal (get-fact.res (fact-simul (+ x 1) st))
                  (factorial x))))
```

It means that any simulation of fact beginning in a initial state st where the current value of arg is a natural x leads (in $x + 1$ steps) to a state where the current value of res is the factorial of x. This theorem is a property of *all* simulations of this design, for all positive values of x.

Therefore, it covers an infinite number of cases, which cannot be done with ordinary simulation.

ACL2 can automatically prove simple properties, but, for more complex properties, it must be guided through previously proved appropriate lemmas. To simplify proofs, the ACL2 model of a VHDL design includes several useful theorems about the design, describing properties of states and of the design representation. As we accumulate experiments, we add more theorems to this standard base, further simplifying the proofs. However, it is the job of the verification engineer to find the right intermediate theorems to guide the tool in a large proof. Usually, these theorems describe the internal behavior of the system, and they are not difficult to find, after some good training on the theorem prover behavior, and on the ACL2 model of VHDL.

An alternative to formal verification is symbolic simulation: we "execute" the system with symbolic values as inputs, obtaining symbolic expressions on the output that describe the system behavior. The input symbols are mathematical variables representing arbitrary values (possibly constrained by some conditions), so one symbolic simulation run can stand for a large or even infinite number of cases. The symbolic simulation is a fully automated process, so it is easier for a system engineer.

With ACL2, we symbolically simulate VHDL designs using the same model we used for simulation and verification. Following Moore [10], we prove the theorem below about our design model. This theorem says that simulating the design 0 times does not change the state. It also says that simulating $n + 1$ times is equal to performing one execution cycle of the design followed by simulating it n times. This theorem is proved automatically, and the hint only optimizes its proof.

```
(defthm unfold-fact-simul
  (and (equal (fact-simul 0 st) st)
       (implies (and (integerp n) (>= n 0))
         (equal (fact-simul (+ n 1) st)
                          (fact-simul n (fact-cycle st)))))
  :hints (("Goal" :in-theory (disable fact-cycle))))
```

The effect of the theorem above is to "program" the theorem prover engine to unfold the definition of the design simulation function when the number of simulation steps is a constant. This, combined with other already available simplification rules, allows ACL2 to simplify calls to the simulation function, which corresponds to performing the symbolic simulation of a design. Thus, in ACL2, symbolic simulation is just a particular case of theorem proving, and all the powerful methods used by this system to simplify expressions are available during symbolic simulation. As a consequence, ACL2 performs very well at the symbol manipulation level when compared to other symbolic simulators.

There are several methods to ask ACL2 to perform the symbolic simulation of a design. The most elementary method proposes a fake theorem in the form (equal (fact-simul *k* *s*) v), where constant *k* is the number of simulation steps, *s* is a symbolic expression describing the initial state, and v is a new variable. We may add assumptions to this theorem, that restrict the initial state *s*. This proposed theorem usually cannot be proved. However, while attempting to prove it, the theorem prover applies the above theorem unfold-fact-simul as a rewriting rule, simplifying the goal (fact-simul *k* *s*). This results in a symbolic expression for each state element. The assigned ones are of particular interest, since they hold the result of the computation.

There are some macros and other facilities to further simplify the symbolic simulation in ACL2, making the answer readily available, but we stick to this simple method here because it emphasizes the connection between theorem proving and symbolic simulation. [10] proposes a symbolic spreadsheet to ease the symbolic simulation with ACL2. Such interface will greatly facilitate the symbolic simulation of VHDL designs.

For instance, the theorem below asks for tree steps of the symbolic simulation of the design fact in an input state where the signal arg (first in the state list) holds the value x. Let us assume this value is an integer larger then 5.

```
(defthm simb-simul-fact-1
    (implies (and (integerp x) (> x 5) (equal st (list x 1 0 0 1 0 0 ...)))
        (equal (fact-simul 3 st) v))
    :otf-flg t)
```

The otf-flg directs the theorem prover to fully expand all proof cases, even the failed ones. Looking at the theorem prover output while attempting to prove this theorem, we find that the expression above was simplified to the following state:

```
(x 1 (* x (- x 1)) 0 1 (- x 2) 1 ...)
```

It means that, after three simulation steps, the current value of signal res (third in the design state) is (* x (- x 1)), and the current value of signal ct (sixth in the design state) is (- x 2). Clearly, this is the beginning of the computation of the factorial of a number x > 5. Please realize that ACL2 uses the hypothesis that x ¿ 5 to decide which branch to take in the conditional statement (Figure 3):

```
WHEN 1 => IF ct > 0 THEN
            res <= res * ct; ct <= ct - 1;
        ELSE
            done <= '1'; state <= 2; END IF;
```

This hypothesis ensures, during the three simulation steps, that signal ct is always greater than zero. For another example, consider the theorem below.

```
(defthm simb-simul-fact-2
    (implies (and (integerp x) (equal st (list x 1 0 0 1 0 0 ...))) ; hyp
        (equal (fact-simul 3 st) v))
    :otf-flg t)
```

It is similar to the previous one, except that we do not assume x ¿ 5. Without this assumption, the prover cannot decide which branch to take in the conditional statement discussed above. This proof attempt generates three cases:

- For the first case, the simplified form is the state (0 1 1 0 1 0 1 ...), meaning that when **arg** (the first state component) is zero, the simulation finishes in a state where **res** (the third state component) is 1.

- In the second case, the simplified form is the state (1 1 1 0 1 0 1 ...), indicating that when **arg** is one, simulation also finishes with **res** equal to one.

- In the the third case, the simplified form is the following state:

```
(implies (> x 1) ; hyp
        (equal v (list x 1 (* x (- x 1)) 0 1 (- x 2) 1 ...)))
```

The hypothesis in this formula indicates $x > 1$. In this case, the simplified state is identical to the first example above, meaning that, after three simulation steps, the simulation reaches an intermediate state where **res** contains (* x (- x 1)). Actually, this is the beginning of the computation of factorial for any $x > 1$.

Since the same tool and the same model are used to perform symbolic simulation and formal verification, the processes can be intertwined, which benefits the two tasks. We use symbolic simulation to find intermediate theorems necessary to large proofs, and as a rewriting rule in other proofs. Conversely, previously proved theorems may be used during symbolic simulation to further simplify the result expressions or reduce the number of generated cases. This is specially useful when dealing with loops.

The tool and the method were applied with success to some simple examples. We proved the correctness of a design for the factorial that uses two processes, one for the state machine, and another for the multiplication unit (sketched in Figure 1). A first proof, using standard ACL2 methods, required about 50 lemmas. A second proof, made with the aid of symbolic verification techniques, required less than 10 lemmas. For another example, consider the two code sequences below:

```
mem[adr1] := a;        x := mem[adr2];
x := mem[adr2];        mem[adr1] := a;
res := x + b;          IF adr1 = adr2
                         THEN res := a + b;
                         ELSE res := x + b;
```

Using symbolic simulation, we find the input-output behavior of these code segments, and using the theorem prover, we show that they are equivalent. These proofs need no engineer guidance. This example was proposed by Gerd Ritter, who was unable to show this equivalence in his symbolic simulator [11]. Another example was proposed by Julia Dushina [5] to decide the equivalence between two high-level state machines, while verifying the correctness of the scheduling step of a high level synthesis tool.

Instr1	Instr2	Instr3
PROCESS BEGIN	PROCESS BEGIN	PROCESS BEGIN
var1 := var2 - in2;	var1 := in1 * var2;	var1 := (in1 + in3) - in2;
var2 := var1 * in3;	var2 := in1 + in3;	var2 := in3 * (in1 * var2);
out1 := in3 + in2;	out1 := in2;	out1 := in3 + in2;
END PROCESS	END PROCESS	END PROCESS

Using a similar method, we show that the sequential composition of sets Instr1 et Instr2 is equivalent to set Instr3.

3.   CONCLUSION

Compared to previous works [4, 2], the new contribution of our research is a formal semantic definition of a synthesizable VHDL subset that remains readable, that is executable, and which automatically produces all the basic useful theorems that will ease the formal proof of the functional correctness of the design. By combining symbolic simulation and theorem proving, we aim at providing the verification engineer with a methodology to efficiently insert formal verification in the very early specification stages of a design. By automating the generation of the formal model from VHDL, we relieve the expert from some tedious tasks (translation, basic theorems, etc). We also make it possible to the new user of our proposed method to rapidly get started on simple models.

Currently, we only support designs synchronized by a single clock. Our work is still on-going in several directions: accept more VHDL primitives, allow the communication between clocked and unclocked components, test the method on bigger designs, and enrich the repertoire of basic theorem templates.

1. REFERENCES

[1] D. Borrione and P. Georgelin. *Formal verification of VHDL using VHDL-like ACL2 models.* In Proceedings of FDL'99, pages 105-116, Ecole Normale Supérieure de Lyon, Lyon, France, 1999.

[2] D. Borrione, editor. *Special issue on VHDL semantics.* Formal Methods in System Design, 7(1/2), 1995.

[3] R. S. Boyer and J. S. Moore. *A Computational Logic Handbook.* Academic Press, Boston, 1988.

[4] C. Delgado Kloos and P. T. Breuer, editors. *Formal Semantics for VHDL.* Kluwer, 1995.

[5] J. Dushina. *Vérification Formelle des Résultats de la Synthèse de Haut Niveau.* PhD thesis, Universit'e Joseph Fourier, Grenoble, 1999.

[6] M. J. C. Gordon and T. F. Melham, editors. *Introduction to HOL; A Theorem Proving Environment for Higher Order Logic.* Cambridge University, Cambridge, 1993.

[7] IEEE Synthesis Interoperability W.G. 1076.6. *IEEE Standard VHDL Language Reference Manual,* 1998. http://www.eda.org/siwg.

[8] M. Kaufmann and J. S. Moore. *An industrial strength theorem prover for a logic based on Common Lisp.* IEEE Transactions on Software Engineering, 23(4):203-13, April 1997.

[9] K. C. McMillan. *Symbolic Model Checking.* Kluwer, Boston, 1993.

[10] J.S Moore. *Symbolic simulation: An ACL2 approach.* In FMCAD'98, pages 334-350,1998. LNCS 1522.

[11] G. Ritter et al. *Formal verification of designs with complex control by symbolic simulation.* In CHARME'99, 1999. LNCS 1703.

[12] N. Shankar. *PVS: Combining specification, proof checking and model checking.* In FMCAD'96, pages 257-264, 1996. LNCS 1166.

Functional Verification with Embedded Checkers

Scott Switzer[1], David Landoll[2] and Thomas Anderson[2]
SmartSand, Inc.[1] : 0-In Design Automation, Inc.[2]

Key words: Functional verification, semiformal verification, embedded checkers

Abstract: Functional verification is now dominating the design cycle for many complex
             chips. Earlier and more thorough detection of bugs is critical for time-to-
             market improvement. One technique that can help is supplementing traditional
             simulation with embedded checkers that monitor for correct design intent
             throughout both block-level and chip-level verification. This paper discusses
             the use of embedded checkers to assist in the functional verification of a dual-
             CPU PCI bridge case study design.

1.   INTRODUCTION

The traditional "black-box" approach to functional verification, in which a verification engineer tests a chip design by developing a testbench to stimulate the inputs and monitor the outputs, is not sufficient for many large designs. The ability to stimulate bugs deep inside the chip strictly from the inputs and to propagate bug effects to the chip outputs is limited. Solutions are needed to improve bug observability, allowing faster and easier detection of bugs, and design controllability, enabling stimulation of corner-case conditions hard to reach with hand-written test cases or random tests.

Embedded checkers that monitor for correct behaviour are one method for improving both observability and controllability. SmartSand and 0-In[TM] Design Automation entered into a joint project to investigate the value of embedded checkers and related technology in the verification of a case study design. SmartSand was interested in new methodologies to accelerate functional verification projects in its ASIC design services business. 0-In wanted to assess the usefulness of its tools on a project that emulated

71

P.J. Ashenden et al. (eds.), System-on-Chip Methodologies & Design Languages, 71–80.
© 2001 *Kluwer Academic Publishers. Printed in the Netherlands.*

customer designs while allowing detailed monitoring of how and where bugs were found.

2.   THE CASE STUDY DESIGN

The case study design project is a dual-CPU PCI bridge that was designed by a team of consultants from SmartSand using embedded checkers from 0-In as part of the verification process.   The design has a high-performance CPU bus that supports split request/data transactions with up to four outstanding transactions at a time.   As shown in Figure 1, the PCI bridge consists of three major blocks: an arbiter that arbitrates on a first-come-first-served (or queue) basis, a CPU interface that handles transactions from the two CPUs, and a PCI master that sends the CPU transactions to the PCI bus.   The team also designed a small PCI slave block that allows access to RAM.   The PCI master and slave implement a subset of the full PCI specification.

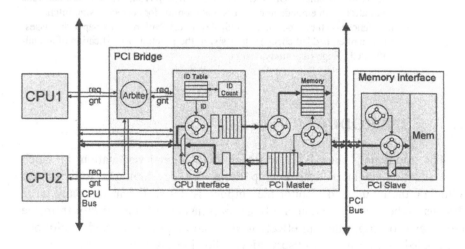

Figure 1. Dual-CPU PCI Bridge Design

The total design was about 50K gates, implemented using synthesizable Verilog RTL for the design itself and behavioural Verilog for the verification testbenches.   The overall development effort consisted of two designers working for three weeks to create and verify all blocks.   An additional engineer was added during the last one and a half weeks of the project for integration and system verification.   Block-level verification involved inserting embedded checkers and writing minimal tests. System verification consisted of a series of directed tests, as well as a random environment.   The embedded checkers within individual blocks were retained in the system verification environment and top-level checkers were added.

3.   BLOCK-LEVEL VERIFICATION

The process of adding embedded checkers to the simulations of the major blocks in the design was a key component of the block-level verification process. The 0-In Check tool recognises "//0in" directive comments in the Verilog RTL source file or in a separate control file that specify which types of checkers are to be added. The checkers are selected from the CheckerWare™ library of more than 50 checkers for data paths, control structures and interfaces. 0-In Check reads in the source files, reads the directives, connects the appropriate checkers and writes out additional Verilog files to be included in the simulation.

Instrumenting the design with checker directives is largely a matter of thinking about what behaviour in the design is legal or illegal. A checker captures design intent by specifying any condition that is illegal and therefore indicates the presence of a design bug. If the design intent is violated in any simulation, the checker "fires" and reports the violation in the Verilog output.

The three main blocks designed and instrumented at the block level were the PCI slave, the arbiter and the CPU interface. The PCI master was largely created through copying the PCI slave, so no significant block-level verification was performed on the PCI master.

4.   PCI SLAVE VERIFICATION

The PCI slave was the first block designed on this project. The block was instrumented with "window" checkers that watch for correct multi-cycle signal relationships and "known/driven" checkers that watch for incorrect X/Z values. These checkers were embedded into the HDL with directives before simulation, and thus found design problems even during the first simulation runs. Training the PCI slave designer on the available checkers and instrumenting the block took 3-4 hours. Many of the bugs found in the PCI slave were detected by embedded checkers rather than by the block-level tests. The designer estimated that the verification cycle for this block was accelerated by 3x from his traditional method of manually examining simulation waveforms (nine days expected; three days actual).

As an example of the problems detected by instrumenting the PCI slave with 0-In Check, one of the window checkers fired when the PCI slave responded to a bus transaction without asserting the "Devsel" signal, a violation of the PCI protocol. This rule was specified in the PCI slave RTL by the checker directive "// 0in awin –start Frame –stop_count 3 –in Devsel" which specifies that "Devsel" must be asserted within three cycles after the "Frame" signal is asserted to mark the start of a bus transaction. Similar checker directives were used to specify additional PCI protocol rules.

This bug was detected when the designer ran a minimal block-level test that stimulated the PCI slave. When the checker fired, the designer understood the bug directly from the firing message, but looked at simulation waveforms to confirm his suspicions that the bug source was a slave state machine problem.

5.   ARBITER VERIFICATION

The arbiter block was very easy to instrument, as 0-In has an "arbiter" checker that reports any violations of a queue arbitration order. As a result, this block required only two checkers for full instrumentation, occupying about 15 minutes of designer time.

Due to the high number of corner cases for the queue arbitration scheme, the designer felt that creating even a moderate block-level testbench would be very time-consuming. She created a minimal test that simply asserted each request line in turn, consisting of only 60 clock cycles. The arbiter block was then simulated with the embedded checkers. Two problems were found during simulation. One of these was a sub-cycle glitch, found by examining the waveforms. The other bug, in which an acknowledge was being generated out of sequence, was detected with the embedded checkers.

To leverage the embedded checkers and minimal testbench, SmartSand decided to try the 0-In Search tool. This uses "semiformal" technology that examines a design and testbench and creates additional stimulus in an attempt to find new ways to fire the embedded checkers. It does this by using the existing tests as a "seed" and then employing formal verification techniques.

0-In Search found five additional functional bugs in the arbiter; no additional bugs were found in this block throughout the remainder of the project. As 0-In Search discovered each bug, it reported a firing and provided scripts to reproduce the test in a Verilog simulator.

Figure 2 shows the simulation waveform for one of the arbiter bugs discovered by 0-In Search. The waveform is divided into two sections; the cycles on the left side of the vertical line are from the testbench, while on the right are new cycles generated by 0-In Search as it took over manipulations of the inputs to set up the conditions to fire the arbiter checker.

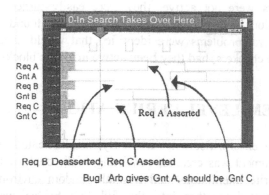

Figure 2. Arbiter bug found by 0-In Search

As shown in Figure 2, this particular bug was a violation of the intended "first-come-first-served" arbitration scheme. The small testbench asserted "req B" and the arbiter responded with a "gnt B" assertion. Then Search took over, dropping "req B" and asserting "req C" simultaneously. One cycle later, Search issued "req A" and then the arbiter issued "gnt A" which was incorrect. Channel C had the oldest request and should have been the next to be granted. 0-In Search discovered this sequence by targeting the arbiter checker with formal analysis to find new ways of firing the checker.

For this block, the value of embedded checkers was enhanced by 0-In Search. Without this additional technology, the designer would have needed a fairly comprehensive block-level test scheme consisting of directed tests and some type of random environment. Being able to simply run 0-In Search and have it create the additional tests necessary was a great help. The arbiter designer estimated that the use of embedded checkers accelerated her verification cycle by 5x (ten days expected; two days actual).

6. CPU INTERFACE VERIFICATION

The CPU interface block was largely designed without the use of embedded checkers. Although this was not intentional, it ended up serving as a comparison of design with and without embedded checkers. Initially, three checkers were added into the CPU interface block, requiring 20 minutes. However, due to a script modification error, the checkers were not active during most of block-level simulation. In fact, the CPU interface passed its block-level tests only because the checkers were not turned on.

Most of the bugs found during system-level verification were due to problems in the CPU interface block. After the engineers realised that the embedded checkers were not active, they were easily turned back on by fixing the simulation script. The bugs that had been found earlier were reviewed, and several problems were identified that would have been more easily found by the checkers, had they been active in all the block-level tests.

7.   SYSTEM-LEVEL VERIFICATION

After block-level verification, a fairly sophisticated system-level verification environment was created, consisting of directed tests for various CPU read and write interactions and a small random environment. The blocks were integrated together into the full PCI bridge system design, including the existing intra-block checkers, and a few additional checkers were added at the top level. The overall system integration and verification effort took three engineer-weeks (one engineer-week creating the environment and two debugging problems).

As previously noted, most bugs found at the system integration level were related to the CPU interface block. The engineers believed that more effective use of checkers in this block would have eliminated one engineer-week from the system verification effort and several days from the block-level verification effort.

After the system verification was believed to be correct, the engineers ran 0-In Search on the complete design. The tool discovered one new corner-case problem that was not detected by the system-level simulations or by block-level 0-In Search runs. This is an especially interesting bug because it arose from a legal scenario that had never been considered by either the PCI master designer or the system-level verification engineer. Therefore, there was no logic to handle this case or any test to check this scenario.

This problem centred on the FIFO in the CPU interface block. The sequence that caused the problem was a CPU two-cycle write, interrupted halfway through. The correct behaviour should have been:
1. CPU1 requests the CPU bus
2. Arbiter grants the CPU bus
3. CPU1 places a write instruction on the bus
4. CPU2 requests CPU bus
5. CPU1 places first data byte on the bus
6. Arbiter drops CPU1 grant and grants bus to CPU2
7. CPU1 stops write after one word
8. CPU interface begins processing first word of two-cycle write
9. First word of two-cycle write is passed to PCI master interface
10. PCI master interface waits for second word of data before processing the full two-cycle write

The bug in the PCI master block occurred in the last step of this sequence. The PCI master recognised that it was receiving a two-cycle write from the CPU interface, but failed to wait for the second word of data. The designer did not include logic to handle an interrupted two-cycle write, so the PCI master simply dequeued two words from the FIFO that supplies data from the CPU interface. Since the FIFO contained only one word of data, the second dequeue was invalid and resulted in bad data being written by the PCI master into the PCI slave memory.

This bug was discovered by 0-In Search when running on the complete integrated design and using the system-level tests as a seed. To ensure that 0-In Search would generate only legal stimulus for the design, input checkers were converted into input constraints. An instrumented design, such as the PCI bridge, includes checkers on major interfaces to monitor for correct protocol during simulation. 0-In Search allows the checkers already on interfaces to be converted into constraints without requiring a special constraint language.

The CPU interface block designer included the 0-In checker directive "//0in fifo –enq push –deq pci_pop –depth 32" in the RTL to identify the location for a checker that would fire if the FIFO was ever enqueued when already full or dequeued when empty. 0-In Search found a way to fire the FIFO "dequeue" check and thereby reveal the bug within an hour. The designer then looked at the checker firing simulation waveforms showing the stimulus that caused the bug. After a brief review, the designer understood the bug and realised that there was no provision for the legal case of an interrupted two-cycle write.

The fix entailed adding new states and logic to the PCI master state machine to handle this scenario. Figure 3 shows the location of the state machine that contained the bug, as well as the FIFO whose checker detected the bug when the FIFO was illegally dequeued. Because of the FIFO checker, the bug was detected close to its source and minimal diagnosis was required to understand and fix the problem.

0-In Search discovered the sequence to fire the FIFO checker even though the verification engineer had never written any tests for two-cycle writes, either interrupted or uninterrupted. The ability to discover new types of legal behaviour that were never exercised in any test is the most important capability of 0-In Search. In the case of this particular bug, the PCI master designer and the system-level verification engineer said that it was unlikely that they would have found this bug without 0-In Search. The entire design was set up for 0-In Search, and this corner-case bug was found, within two engineer-days.

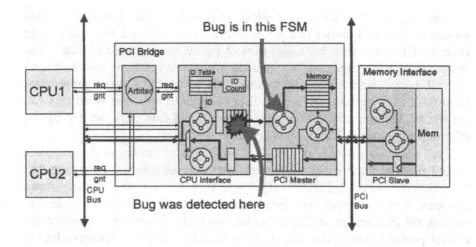

Figure 3. System-level bug found by 0-In Search

8. BUG RESULTS

Embedded checkers found the majority of the bugs in this case study design. Specifically, checkers found 17 of 19 PCI slave bugs (89%) at the block level and 6 of 7 bugs (86%) in the arbiter. Even in the CPU interface block, with its limited use of checkers, 6 of 17 bugs (35%) were found via checkers. At the system level, with all blocks integrated together, 7 of 12 bugs (58%) were found with embedded checkers.

In total, 55 bugs were found in the design. Of these, 43 bugs (79%) were found at the block level and 12 bugs (21%) were found at the system level. This was a desirable side effect of using embedded checkers at the block level, as bugs are easier to diagnose and fix at the block level. In general, moving verification to the block level accelerates the overall verification cycle. Figure 4 shows a summary of bugs found throughout the project.

Figure 5 summarises the location and method of bug detection. Of the 55 total bugs, embedded checkers found 36 (65%). Of these 36 bugs, embedded checkers stimulated by the testbench caught 23 and embedded checkers stimulated by 0-In Search caught 9. The testbench and checkers in simulation simultaneously detected 4 bugs. These bugs were easier to debug with the embedded checkers, but would have been caught without the checkers. Of the 19 bugs (35%) not found by embedded checkers, the testbench was able to catch 6 through comparison of expected and actual data. Other detection schemes, including visual inspection of simulation waveforms and code reviews, detected the remaining 13 bugs.

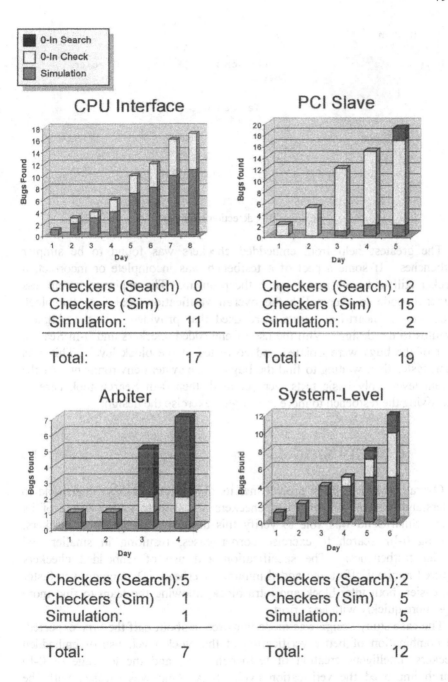

Figure 4. Block-level and system-level bug statistics

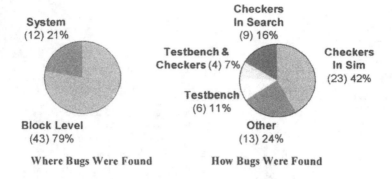

Figure 5. Bug detection summary

The greatest help from embedded checkers was found to be simpler testbenches. If some aspect of a testbench was incomplete or incorrect, a checker still had a chance to catch the problem. The engineers wrote less testbench code to accomplish the overall verification goal. At the block level, simple "heart-beat" tests were used that provided only the simplest stimulus to the design. With the use of embedded checkers and 0-In Search, most of the bugs were isolated and corrected at the block level. This was much faster than waiting to find the bugs in the system environment. At the system level, only basic tests were created, then 0-In Search took care of amplifying the testbench to more thoroughly exercise the design.

9.   CONCLUSION

Overall, the case study project met its objectives to allow SmartSand to understand the role of embedded checkers and 0-In to assess usage of its tools. SmartSand was able to verify this design with embedded checkers, utilising 0-In Search to exercise corner cases, resulting in smaller and simpler testbenches. The specification and use of embedded checkers required minimal effort and added minimal overhead. Debugging was faster and easier, both inter-block and intra-block, allowing the team to find more bugs more quickly with less effort.

The case study design was made bug-free in about half the time expected; the combination of better verification at the block level, use of embedded checkers, intelligent creation of testbench tests, and the leverage of 0-In Search improved the verification cycle. SmartSand was pleased with the outcome of this design and verification process and continues to use the 0-In tools when applicable for new projects.

Improved Design Verification by Random Simulation Guided by Genetic Algorithms

Pierre Faye, Eduard Cerny
Dept. IRO, Université de Montréal, Montréal (Québec) Canada. {faye, cerny}@iro.umontreal.ca

Keywords: H/W Design Verification, Guided Simulation, Genetic Algorithms

Abstract: We describe a method to increase design verification coverage by guiding simulation using random input sequences. The simulation control algorithm is inspired by genetic optimization algorithms. The optimization criteria are the activation counts of various property checkers, each being concerned with a specific function of the design. Experimental results on an RTL Verilog design of some 6000 equivalent gates show that the same property coverage as in pure random simulation with uniform distributions can be achieved with half the number of simulation vectors, or, conversely, with the same number of input vectors the activation count is roughly doubled. Similar results were obtained on a 800 K gate design

1. INTRODUCTION

The amount of Verilog or VHDL RTL code of a hardware system is large, but the code of the simulation test benches to verify the system and component models is much larger and growing [1]. This is due to the desire to improve the quality of the design and the growing size of the chips to be verified. Many new techniques and tools are reported in the literature and some appear as commercial tools: Improved verification productivity through better test-bench construction environments [e.g., 8, 9]. Reduced

P.J. Ashenden et al. (eds.), System-on-Chip Methodologies & Design Languages, 81–95.

number of testbench cases by using random input sequences [8, 9]. Improved observability through property checkers [6]. Improved coverage evaluation through software code-coverage techniques [10] or based on state / transition information of the design [6]. Symbolic model checking of a temporal formulas [14, 5] or using language inclusion tests of w-automata [4]. Partial state-space exploration have been developed [7]. Theorem proving at abstract design level has also been used [4, 15].

Since formal techniques still cannot handle large designs, simulation remains the main means of RTL design verification. The size of the testbenches is determined by the number of different features that should be verified, and by the complexity of the input sequences. The problem of developing the appropriate input sequences is the most difficult part.

An approach that reduces the size and the number of the test benches is to use simpler dedicated sequences to remove trivial design errors, and then to use random input sequences to explore the rest of the input space. A number of simulation environments exist that help to carry out random simulations [6, 8, 9]. The problem is that even with long random sequences there is no guarantee that important corner cases will in fact be addressed. It depends very much on the random distributions, the importance of different input ports, and the complexity of the design. The question arises whether it is possible to guide the random simulation in some way to achieve better coverage. Clearly, knowledge of the design structure helps to select better random distributions, but if the outcome is not good enough, then it may not be obvious how to adjust the parameters.

In this paper we report on influencing the random distribution parameters to improve property coverage using a method inspired by genetic algorithms [20]. Our method can be summarized as follows: A population of short simulations is carried out, each using a different set of parameters for the random input generators. Verification coverage targets are updated as a result of these simulations. The weakest target is selected and each sequence is evaluated as to its contribution to improving that target. This assigns a fitness value to each member of the population. Genetic crossover and mutation operations are applied to the distribution parameter vectors, and a new population of short simulations is carried out, while updating all the coverage targets. This is repeated until that target reaches above a user specified threshold. Then the next weakest coverage target is selected for improvement, etc., until all targets reach the predetermined limit (threshold) or the specified max. CPU time is used up.

Genetic algorithms have been used to optimize complex objective functions. The convergence is generally slow, but the advantage is that it can deal with large problems. In design verification, there are not yet many applications of this technique. Some of the reported applications of genetic algorithms in HW/SW design dealt with the verification of communication protocols under traffic stress conditions [18, 19] and with test pattern generation [16, 17].

We carried out an experiment on a design inspired by an industrial application, consisting of some 6000 equivalent gates in synthesizable RTL Verilog. The lengths of a sequence to reach a fixed coverage level using our method (the total number of input vectors from all individual simulations over all populations) is about half the number of vectors required on the average (over a number of simulations with different seeds) by a pure random simulation with uniform distributions on all ports. Similarly, for a fixed sequence length, our method doubles the activation counts. Later, we also verified a number of properties on a 800K-gate design and obtained similar results.

The paper is organized as follows: In Section 2 we describe our verification method, in Section 3 we present experimental results, and in Section 4 we conclude the presentation.

2. THE VERIFICATION ENVIRONMENT

Given an RTL design, we construct a model of the surrounding circuitry (model of the user environment) so as to allow non-deterministic choices in the construction of any valid input sequence. These choices are controlled from a set of primary inputs of integer or enumerated type. We then specify the type of the distribution of values on these primary input ports (and possibly their mutual correlations), thus indicating to the system how to form random input sequences using a random number generator. This forms a *parameter* vector of the simulation. Each simulation sequence of length L is thus characterized by the initial seed of the generator and the parameter vector: the *chromosome*.

2.1 Genetic algorithms

Genetic algorithms are based on the Darwinian theory of natural selection. Initially, a population of individuals (chromosomes) is randomly created (or may be supplied by the user). A fitness value is then attributed to each individual of the population, giving a measure of quality according to a fitness function. After the whole population has been evaluated, genetic operations, like cross-over and mutation, are performed over the best individuals to create a new population. This process (Figure 2) is repeated over a number of generations until a certain level of quality of the population or the max. simulation time is reached.

2.2   Encoding and implementation

- Chromosome Encoding:

A chromosome (individual) represents a simulation sequence of length L. It is characterized by L and a vector of integers that characterize the distributions of values on the n primary input ports.

- Fitness Evaluation:

The coverage information of the design could be obtained from various sources [10, 6], but for this experiment we defined RTL coverage through a set of properties (initially given in CTL [5]) of the form *precondition -> conclusion,* covering different functional aspects of the design. The property is converted to a VHDL or Verilog passive process that observes the behavior of the design and the environment (e.g., as in [21]). The coverage consists of the activation count of a property or group of related properties. An activation count is the number of times the precondition of the property became true during a simulatio. Then,

> Fitness value of an individual simulation = Activation count of the property group selected during the simulation

- Selection and Crossover Scheme:

A population consists of N simulations. The population is ordered by the fitness values, and the epuration factor (epf) portion of the population with low fitness values is dropped. From the remaining individuals, two are randomly selected for crossbreeding. This random selection of chromosomes is weighted by the individuals' fitness value, in order that highly fit individuals have a better chance to contribute to the next generation. For simplicity, the crossover operation (Figure 2) generates two children, each composed of 50% of the distribution parameter vector from one parent and 50% from the other parent.

- Mutation Scheme:

Each child is selected for mutation with a given probability (*mutate*). An element of the chromosome vector (Figure 2) is randomly chosen and is randomly modified. While *crossbreeding* tries to improve fitnes, *mutation* tries to get the sequences out of a local optimum (similar state transitions) into other parts of the inputs sequence space.

- The Pseudo code of the Genetic Algorithm is shown in Fig 1:

```
begin
    new_population = Generate_random_initial_population;
    while (desired level of quality or maximum simulation time is not reached) // Main loop
        old_population = new population;
        new_poplation = {};

        // select the best individuals. This results in a smaller but better population.
        epurated_population = select_best_individual (old_population);

        loop-for (ind = 1 to max population) // simulate and compute fitness of each individual
            simulation_run_with(individual[ind]);
            evaluate_fitness_of(individual[ind]);
        end loop-for;

        repeat until (|new_population| == N) // Apply genetic operators on the population
            (parent1, parent2) = randomly_pick_according_to_fitness (epurated_population);
            (child1, child2) = crossbreed(parent1,parent2); // cross-over
            if (mutate) randomly_chose_a_parameter_and_mutate_it (child1);
            if (mutate) randomly_chose_a_parameter_and_mutate_it (child2);
            new_population = new_population U {child1,child2};
        end repeat
    end while
end§
```

Figure 1. Pseudo code of the Genetic Algorithm

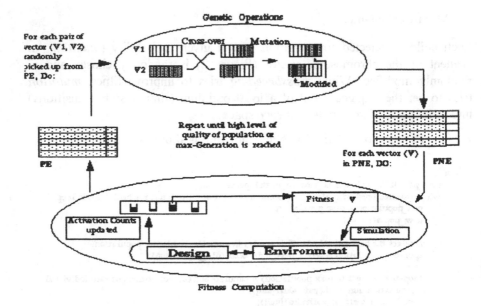

PNE = Population with individuals' fitness value Not Evaluated yet.

PE = Population with individuals' fitness value Evaluated.

Figure 2. Random Simulation Guided by a Genetic Algorithm

The global simulation parameters are:

- Population size N: The number of individual simulation runs (N = 200 was sufficient for the 6K-gate design).

- The number L ofclock cycles of an individual simulation. This is determined by experience from initial directed simulations, it must be long enough to allow the system to react to a number of input stimuli. (In the 6K-gate design L = 600.)

- Threshold Thresh_P: The minimum number of activations of a property group. The entire simulation terminates when all property-group activation counts reach above the threshold or the max CPU time is reached. (In the 6K-gate design Thresh_P = 8000.)

Initialization will set:

- Randomly or by user, the distribution parameters for the N individuals in the initial population.

- Randomly or by user, one of the property groups.

Global simulation control:

- Carry out N individual simulations, each of length L. During each simulation continue updating activation counts of all property groups and record the improvement (Figure 2).

- When the N simulations are completed, if the activation count reaches above the threshold, switch to the property group with the currently lowest activation count as the next objective, otherwise keep the current group. Generate a new population.

- Repeat until either all counts reach above the threshold or max CPU time is reached.

Note that all properties are checked in all individual simulations and their activation counts are also updated, even though only one group is the focus of the current optimization step. All simulation runs thus contribute to the verification of the design.

3. EXPERIMENTAL RESULTS

We report on an experiment carried out on an RTL Verilog model inspired by an industrial Transaction Reordering Subsystem (TRS). We constructed a non-deterministic model of the use environment of the subsystem, such that it could select any valid input. The size of the combined model (Figure 3) is some 6000 equivalent gates. The model was also analyzed with FormalCheck and it reported 2.00e+07 states reached (326 state variables and 3 combinational variables) for Property 1 shown in Section 3.2.

We carried out the following simulation experiments:

1. Perform the optimizing simulations until all counts are above the threshold. Repeat for different initial parameters. Record the mean, min. and max. values of L and CPU times.

2. Perform a number of pure random simulations with different seeds and uniform distributions until all the counts reach above the threshold.

We then compared the efficiency (length of simulation, coverage) of the optimizing simulations with the pure random simulations. We examined the min, max, and average values, and the coverage (activation counts) achieved by random simulations with the same lengths as the min. and max. values of L obtained by using the genetic algorithm.

3.1   The design and environment models

A block diagram of the TRS and its non-deterministic environment model are shown in Figure 3.

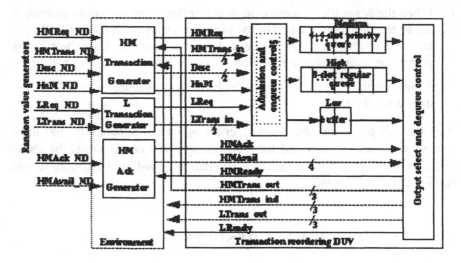

Figure 3. Transaction reordering subsystem and its non-deterministic user

The "user" environment of the design is controlled by the input signals denoted as xxx_ND. Each of them randomly selects the corresponding operation in the current clock cycle. For instance, HMReq_ND selects whether a request is to be made for the Transaction number generated on HMTrans_ND. If that number is not currently busy, then the Generator produces a real request on the input of the TRS, while if that number is busy, no request is generated. The Desc_ND values are directly transferred to the Desc input of TRS.

Each xxx_ND signal is associated with a type of random distribution and its parameters. As these signals refer to different functions of the TRS, the types of distribution associated with them are also different. For instance, the parameter associated with the 1-bit-xxx_ND signals (HMReq_ND, HnM_ND, LReq_ND, HMAck_ND) gives the probability of the random generators corresponding to output 1. For the transaction number generator associated with the 3-bit-HMTrans_ND signal, we assign a probability of occurrence p_i to each transaction number $i = 0, 1, ..., 7$. This results in a set of parameters, each representing the probability of a particular transaction number to be output by HMTrans_ND, with $\sum p_i = 1$.

3.2   Coverage metrics and genetic parameters

The following properties stated here verbally were first expressed in CTL referring to the inputs and outputs of the TRS and the state of an occupation table of emitted requests in the HM Transaction Generator. These were then converted to extended synchronous finite-state machines, expressed as Verilog processes. Each process checks that the corresponding property holds and increments an activation count. The processes are integrated with the rest of the model and the overall simulation control.

3.2.1   Properties:

We selected 4 representative property groups to verify (Table 1). There are many more properties, but their preconditions are similar, thus their activation does not provide more data for analyzing the guidance algorithm.

Table 1. Property groups to verify for the 6K-gate circuit

Groups	Description of the Property Groups
p1	For all $0 \leq i \leq 7$, when a Transaction number i arrives with HMReq = 1 (the precondition), then the occupation table entry i is 1 and gets reset to 0 on the next clock cycle (the conclusion), thus memorizing that this number has been submitted to TRS.
p2	When a Transaction Number i enters the TRS, then before it is output on HMTrans_out and acknowledged, then the same Transaction number is not submitted to TRS with HMReq =1.
p3	For all $0 \leq i \neq j \leq 7$, when a Medium Priority Transaction number i comes into the TRS after a High Priority one j, then the High Priority one must leave the TRS first.
p4	When a Transaction is output on HMTrans_out with HMReady = 1 and it is acknowledged, then the same Transaction number appears on HMTrans_ind 3 clock cycles later.

3.3   Sensitivity analysis of N and L

When we deal with genetic algorithms, it is far from obvious to decide which numerical values to assign to the genetic parameters like the individual simulation length (L), the population size (N), the maximum number of generations (max_gen), etc.

The success of the method depends on a "clever" choice of these parameters. This section presents simulation results that show the value(s) of L and N that optimize the activation counts. For different values of L we run several simulations. The results are summarized in Table 2.

Table 2. Sensitivity Analysis of N and L

L / N	20	50	80	100	150	200	230	250	300
300	1.28	1.45	1.58	2.00	1.78	1.81	2.03	2.01	2.18
600	1.57	1.61	2.24	2.17	2.54	2.72	2.78	3.04	3.04
700	1.76	1.86	2.04	1.74	2.31	2.61	2.78	3.07	2.81
1000	1.64	1.90	2.01	2.65	3.02	3.09	2.75	3.17	2.53

For each entry in Table 2, we ran a pure random simulation followed by a genetically-guided random simulation. The reference used for the genetic simulation is the activation count of the Property 3 (the more difficult one).

For example, consider the entry (150, 600). The populations have N = 150 individuals, each being a simulation run of L = 600 clock cycles (a new population is "genetically" generated after 150 x 600 = 90000 cycles).

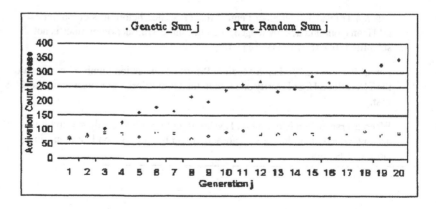

Figure 4. Coverage Improvement by genetic algorithm for N = 150 and L = 600

Figure 4 shows the coverage improvement of the genetic simulation over the random simulation. First we run a genetic simulation (with (150, 600) as the pair of parameters) and for each generation we sum the 150 fitness values. Given a generation j, j ∈ {1..max_gen}   (max_gen = 20 in this example),

Genetic_Sum_j $=\Sigma$ *fitness value of individual (i).* These points (Genetic_Sum_j) are plotted to form the "Genetic" curve of Figure 4. Then we run the pure random simulation (with the same pair of parameters). Each time we applied the same number of input vectors as to the "genetic" population (90000 input vectors), we recorded the contents of the activation counter (corresponding to p3) and then reset this count to 0. This results in a set of points, {Pure_Random_Sum_j, j $\in$ {1..max_gen}}, which are plotted to form the "Pure Random" curve of the Figure 4.

$$\text{Efficiency}(N, L) = \left(\left(\sum_{j=1}^{max\_gen} Genetic\_Sum_j\right) \bigg/ \left(\sum_{j=1}^{max\_gen} Pure\_Random\_Sum_j\right)\right)$$

gives the (N, L) entries of Table 2.

The results presented in Table 2 show that using Property 3 as the reference, the genetic method is more efficient than the pure random one, providing an activation count that is up to 3.17 times higher in only 20 generations.

An analysis of the mean value of Efficiency(N, L) over the different values of L shows that for a fixed value of N, Efficiency(N, L) tends to increases with L. We can (empirically) conclude that the experiments using 300 clock cycles (or less) per simulation run, though favoring the genetic method (Efficiency(N, L = 300)) > 1), do not reach a good enough level of efficiency.

We can easily identify one area at the bottom right hand side of the Table 2 where the highest values of the (N, L) entries (> 2.60) are located. Among these "good" (N,L) pairs, we have the choice between using a fewer individual vectors in a population and longer individual simulation runs or using more individual vectors and shorter simulations. The decisive criteria here is the simulation time. As an example, we can notice that (200,600) and (230,1000) have about the same level of efficiency (2.72 vs. 2.75, respectively). The pair (230,1000) leads to a simulation length of 230 x 1000 x 20 = 4,600,000 cycles, while the pair (200,600) requires only 200 x 600 x 20 = 2,400,000 cycles. N = 200 and L = 600 seems to be a good choice for this circuit and this property.

The experiments presented in this section optimized only one property (p3) and for not more than 20 generations. The only purpose of these short experiments was to allow us to make an suitable choice of the genetic parameters (L and N) before executing the main genetic experiment (using the methodology in Section 2.2). The results are reported in the next section.

3.4   Simulation results

We executed 10 simulations guided by the genetic algorithm and 10 pure random simulations, each with a different seed, on a Sun ULTRA SPARC 1 Station with 252 MB of memory. The guided simulation required about 7148 seconds (less than 2 hours) of CPU time (average over the 10 experiments), whereas the random one never reached the Activation count thresholds on all property groups within the allotted max. number of simulation vectors, and requiring more than 4 hours of CPU time.

The parameters used by the genetic algorithm are summarized in Table 3

Table 3. Genetic parameters for the 6K-gate circuit

PARAMETER	NAME	VALUE
Individual simulation length [Clock cycles]	L	600
Population size	N	200
Max nb. of Generation	max_gen	45
Range of exponential parameter	param_range	100
Mutation probability	mutate	25%
Property group Activation count threshold	Thresh_P	8000

In what follows:

- I is the simulation index number,

- T is the CPU time required by the simulations [seconds],

- V is the number of individual vectors (chromosomes) required in the guided simulations. The product V x L gives (clock cycles). In the pure random simulations, V is the number of input vectors divided by L. (The division by L allowscomparing the two forms of simulation.)

- Ny is the activation count of property group $y = 1,..., 4$.

- The maximum number of individual vectors for pure random simulations was set to 9000 (corresponding to 1,800,000 clock cycles).

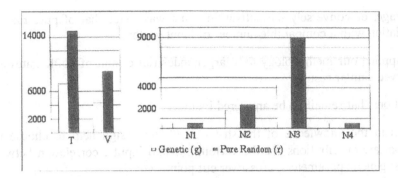

Figure 5. Comparison of Tx, Vx, Nyx (mean over 10 experiments) between Genetic and Random Simulations

Figure 5 shows that to reach the same level of Activation counts, the pure random simulation needed at least $L = 9000$ input vectors which is roughly 2x as much as the guided simulations. Similarly for the CPU time (T). For example, to reach the same level of coverage on a large design, the guided simulation could run only for 3 days instead of 6.

After the initial random choice, the genetic algorithm always selected the more difficult property to activate (p3), and then adjusted the parameter vector of the means for the distributions so as to increase its activation count faster than the pure random simulation.

If the same number of vectors were applied in the random simulation as in the guided simulation of the more difficult property group (p3), then the Activation count would not have been even 1/2 of the threshold of 8000.

4. CONCLUSIONS

We have described a method inspired by genetic algorithms to perform "guided" random simulation with the objective to increase the design verification coverage as compared to a pure random simulation. We reported on an experiment in which we applied our method to an RTL Verilog design consisting of some 6000 equivalent gates. The coverage metric consisted of the values of Activation counts of four property groups implemented as finite state machine checkers (passive Verilog processes). The results were encouraging in that it demonstrated that it is possible to influence the generation of random input sequences so as to increase the counts more rapidly than using a pure random simulation. In the presence of a more difficult property (p3), the simulation time was cut in half to reach the same coverage, or conversely the activation count was twice that of pure random simulation with a comparable number of input vectors.

We applied our methodology to a larger industrial circuit of 800K gates and observed similar results.

Questions that remain to be answered include:

How can the knowledge of the structure of the design help to choose the appropriate distributions for the various random inputs, correlation between inputs, genetic parameters, crossover cut points, etc.

What are the appropriate coverage metrics and how to group them. Could one use code coverage [10] or transition-pair coverage [6] ?

In the case that one coverage parameter (e.g., one activation count) is never activated, how to guide the input generators to target this parameter? Could user-supplied proximity metrics between the property groups be part of the fitness function for evaluating individuals?

How to detect that the increase in the activation count of the currently optimized property is leveling off and it is thus possible to switch to another property before reaching the max. number of generations?

REFERENCES:

[1] A. Silburt, "Invited Lecture: ASIC/System Hardware Verification at Nortel: A View from the Trenches", Proceeding of the 9th IFIP WG10.5 Advanced Research Working Conference on Correct Hardware Design and Verification Methods (CHARME'97), Montreal, October, 1997.

[2] A.J. Camilleri. "A role for Theorem Proving in Multi-processor Design", Proceeding of the 10th International Conference on Computer Aided Verification (CAV'98), Vancouver, June/July 1998.

[3] D.L.Dill, "What's Between Simulation and Formal Verification", Proceeding of the 35th Design Automation Conference (DAC'98), San Francisco, CA, USA, June 1998.

[4] FormalCheck User's Guide. Bell labs Design Automation, Lucent Technologies, V1.1, 1997

[5] K. L. McMillan, "Symbolic model checking - an approach to the state explosion problem", Ph.D. thesis, SCS, Carnegie Mellon University, 1992. (Also Cadence Design Systems, www pages.)

[6] 0-In Design Automation, Inc., "Whitebox verification," http://www.0-in.com/tools.

[7] 0-In Design Automation, Inc., "0-In search," http://www.0-in.com/tools.

[8] System Science, Inc. (Synopsys, Inc.), "The VERA verification system," http://www.systems.com/products/vera.

[9] Verisity Design, Inc., "Spec-man: Spec-based approach to automate functional verification," http://www.verisity.com.

[10] TransEDA, Inc., "VeriSure- Verilog code coverage tool," http://www.transeda.com/products.

[11] Lucent Technologies,"FormalCheck model checker," http:/www.bell-labs.com/org/blda/product.formal.html.

[12] P. Bratley, B.L. Fox, L.E. Schrage, A Guide to Simulation, 2nd Edition, Springer-Verlag, New York, 1986.

[13] L. Devroye, Non-Uniform Random Variate Generation, Springer-Verlag, New York, 1986.

[14] CheckOff User Guide, Siemens Nixdorf Informations Systemen AG & Abstract Hardware Limited, January, 1996.

[15] Lambda user's manual, Abstract Hardware Limited, 1995.

[16] E. M. Rudnick, G.S. Greenstein, "A Genetic Algorithm Framework for Test Generation", IEEE Transaction on Computer-Aided Design of Integrated Circuits and System. Vol. 1, No. 9. September 1997

[17] F. Corno, P. Prinetto, M. Rebaudengo, M. Sonza Reorda. "GATTO: A Genetic Algorithm for

[18] Automatic Test Generation for Large Synchronous Sequential Circuits". IEEE Transaction on Computer-Aided Design Networking, Vol. CAD-15, No. 8, August 1996.

[19] M. Baldi, F. Corno, M. Rebaudengo, P. Prinetto, M. Sonza Reorda, G. Squillero, "Verification of Correctness and Performance of Network Protocols via Simulation-Based Techniques", Conference on Correct Hardware Design and Verification Methods (CHARME'97), Montreal, Canada, October 1997.

[20] E. Alba, J. M. Troya, "Genetic Algorithms for Protocol Validation", Proceedings of the 4th conference on Parallel Problem Solving (PPSN IV), Berlin, Germany, September 1996.

[21] Goldberg, D. E., Genetic Algorithms, in Search, Optimization and Machine Learning, Addison-Wesley, 1989.

[22] W. Canfield, E. A. Emerson, A. Saha., "Checking Formal Specifications under Simulation", International Conference on Computer Design (ICCD'1997), Austin, Texas, October 1997.

VERIS: An Efficient Model Checker for Synchronous VHDL Designs[*]

Yiping Fan, Jinsong Bei, Jinian Bian, Hongxi Xue, Xianlong Hong, Jun Gu[†]
Email: {fanyp | beijs | bianjn | xuehx | hong}@tiger.cs.tsinghua.edu.cn
Email: gu@cu.ust.hk
Dept. of Compute[†] Dept. of Computer Science, Hong Kong University of Science and Technology, Hong Kong, China r Science and Technology, Tsinghua University, Beijing 100084, China

Key words: Formal Verification, Model Checker, VHDL, Synchronous Circuits, FSM

Abstract: In this paper a solution for property verification of synchronous VHDL designs is introduced, and an efficient symbolic model checker is implemented. The model checker applies the feature of synchronous circuit design and the locality feature of property to reduce the size of the state space of the internal finite state machine (FSM) model, thus speeding up the reachability analysis and property checking of circuits. A counterexample generation mechanism is also implemented. We have used the implemented model checker to verify several benchmark circuits; the experimental results contrast with another well-known model checker and demonstrate that our solution is more practicable.

1. Introduction

Interest in formal verification techniques for hardware designs has been growing recently years. The most effective and successful formal technique is model checking [1, 2], which explores the state space of designs to check if it meets some properties.

A lot of subtle research has been introduced to give a formal semantics to VHDL and to apply formal verification techniques [3, 10, 11]. However, VHDL

[*] This work was supported by Chinese National Key Fundamental Research and Development Plan (973) under Grant No. G1998030404.

P.J. Ashenden et al. (eds.), System-on-Chip Methodologies & Design Languages, 97–107.
© 2001 *Kluwer Academic Publishers. Printed in the Netherlands.*

is a very complex language and the models that capture all the VHDL features are almost inherently not acceptable for design automation tools. So we must trade off the number of VHDL features modeled, and the practical usefulness of the semantics, for achieving more feasibility. D. Deharbe defined the operable semantics of a verification-oriented subset of VHDL, in terms of abstract machine [4]. In this model, state transition models a VHDL delta cycle. However, Deharbe's method is still too complex, and the result of states is unreasonable when it is adopted to represent *synchronous* circuits. Most of the digital systems are synchronous circuits, so we should find an efficient verification method for them.

We have developed a verification system VERIS for synchronous VHDL designs. The verification system is based on symbolic model checking like Deharbe's model. However, our approach exploits the features of synchronous circuits to propose a new FSM model, which results in reducing state spaces dramatically. By removing irrelevant portions of the circuit according to the property to be verified, we can reduce the state space and speed up the verification further. Our results show that the state machine is more reasonable, the size of state space is reduced several orders of magnitude and becomes readable, while unreadable when using Deharbe's mixed modeling method.

To reduce the size of state model further more efficiently, we apply the locality feature of property to eliminate parts of the model that are not relevant to the specification. The model checker needs to consider only the transition functions of the variables in the cone of influence of the specification.

To make the model checker more practicable, we give counterexamples when it does not meet any properties.

In the next section , we give the outline of our model checker. In Section 3, we describe the model elaborator, which converts the synchronous VHDL design to an internal model. Then we describe the property verification in Section 4, including the elimination of irrelevant portion and counterexample generation. In Section 5, we show the experimental results, and finally, some conclusions are shown.

2.   System structure

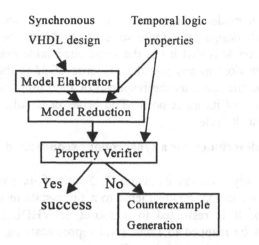

Figure 1, System structure of VERIS

The structure of VERIS is illustrated in Figure 1. We can see that the system consists of 4 main parts:

Model elaborator. It reads the VHDL description of a synchronous circuit design, and then builds the corresponding internal model, which is a symbolic BDD-based finite state model. This step is crucial for the entire verification process since the complexity of the model checking algorithms heavily depends on the size of the model (i.e. the number of state variables, or the size of the BDDs).

Model reduction part. It focuses on the locality feature of property to eliminate parts of the model that are not relevant to specifications.

Property verifier. It reads a specification file composed of temporal logic properties, extracts the relevant cone portion to get the reduced model, then it traverses the reduced symbolic model to determine if the design implementation satisfies the property specification. The temporal logic specification and the VHDL design are physical independent and reside in separate files. Then we can modify the properties to be checked without re-elaborating the VHDL design. On the other hand, in a project design, the same VHDL design may be a component of a larger module. It is also not reasonable to provide the specification and design in a same file.

Counterexample Generation part. When an error is detected, the property verifier can produce a counterexample as a VHDL testbench. This information can then be used to debug the circuit design.

3.   Model elaborator

The role of the model elaborator is as the following: Given a synchronous sequential VHDL design D, a finite state model M can be elaborated. The requirement put on M is that it has the same observable behavior as D. The "Observable behavior" means that the response of M to the stimulus on its inputs should be the same as the response of the output ports of D to the same input values of its input ports. The behavior should be considered at the level of the clock cycle.

3.1   VHDL description for a synchronous sequential design

VHDL is a very complex language [5]. Some of its constructs generate infinite models and cannot be abstracted to a finite-state model, for example, an infinite loop. It is restricted to a subset of VHDL such that design descriptions can be mapped to finite state representations. That is, objects must be of a finite type (no access or file types, no unconstrained arrays, and no generics) and quantitative timing information is not accepted (any *after* clauses in assignment statements, and any *for* clauses in *wait* statements).

We further restrict the circuit design type to synchronous circuits. A synchronous sequential circuit must have a system clock. The inputs are introduced into the circuit to process sequentially, and to generate outputs by the controlling of the system clock. That is, the external events are synchronized with the system clock. If a process or a concurrent statement is independent of the system clock, it must be a combination part. In that case, we can reduce the combination parts in the phase of elaborating the model.

3.2   Synchronized finite state model

According to the synchronous behavior, we can define a model $M = (S, I, O, s_0, TF, OF, clk)$, called *synchronized finite state model* [10], where,
S is a power of B, which represents the states of the machine, $S = B^{ns}$, and

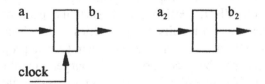

Figure 2, Sequential variable VS.

Combinatorial variable

$s_1, s_2, \ldots, s_{ns}$ are the corresponding state variables.
I is a power of B, which represents the inputs of the machine, $I = B^{ni}$, and $i_1, i_2, \ldots, i_{ni}$ are the corresponding input variables.

O is a power of B, which represents the outputs of the machine, $O = B^{no}$, and $o_1, o_2, ..., o_{no}$ are the corresponding output variables.

$s_0 \in S$, represents the initial state of the machine.

TF: $S \times I \to S$: $TF = \{tf_1, tf_2, ..., tf_{ns}\}$, TF represents the next state function, and tf_i: $S \times I \to B$ is the transition function of the state variable s_i.

OF: $S \times I \to O$: $OF = \{of_1, of_2, ..., of_{ns}\}$, OF represents the output function, and of_i: $S \times I \to B$ is the output function associated to the variable o_i.

clk is the global clock of the synchronized finite state machine.

B denotes the usual Boolean domain ($B = \{TRUE, FALSE\}$).

There is an implicit synchronous clock (clk) for each tf and of, which means states or outputs can only be changed when the events of clk occurs.

3.3   Modeling algorithm

In VHDL manual, process statements are the atomic components of a design entity, and any VHDL concurrent statement except a block statement has a corresponding equivalent process statement. So we can decompose the transformation of any VHDL design unit to the synchronized model by two steps:

First, we need elaborate a sub-model for each VHDL process statements or equivalent concurrent statements in certain declaration environment. The following is the elaboration steps:

Determine the input variable and output variable of the sub-model.

Analysis every *wait* statements and assigned objects. Mark each signal or variable whether the system clock controls it or not.

Resolve data dependencies.

a. Get assignment decision diagram for every assigned object
by deriving an execution tree.

Generate sub-models represented by BDD.

After elaborated sub-models from all process statements or equivalent concurrent statements, the next work is to compose all the sub-models to obtain the final model of the VHDL design entity. The following gives the composition mechanism:

Compose all of sub-models parallelly.

Declaration encapsulation, which means adjusting the input variables and output variables according to the ports and signals of corresponding entity declarations.

Reduce intermediate signals or variables. Reduce signals or variables independent of the system clock by substituting them with their assignment decision diagrams.

These steps can be explained as an example illustrated in figure 2. In the left of figure 2, the signal b_1 is assigned by a_1 and controlled by the system clock also. So we consider b_1 as a sequential variable and model it as an output of some transition function of FSM. While, in the right of figure 2, the output signal b_2 is assigned by a_2 and is independent of clock. So it must be a combinatorial variable and be reduced by replacing it with a_2.

So in the final model, all the state transitions represent the changes of state variables relating to the system clock cycle. All combination parts are abstracted to Boolean expressions and denoted by the transition functions, which allows our models to use fewer variables and fewer bits to represent one VHDL designs. So the final model yielded by above algorithm is dramatically smaller in size than that yielded by Deharbe's method [4], which observably speed up the whole model checking.

4. Property verification

The property verifier accepts the internal model and the property specification as inputs, and traverses the state space of the model to determine whether the property is satisfied.

4.1 Temporal logic specification

The language used for specifying the expected behavior of a VHDL design is essentially the temporal logic CTL (Computation Tree Logic) [6]. A specification is composed of a set of properties about a VHDL description. It is the role of the model checker to verify whether the VHDL description satisfies the properties.

Properties are represented in *computational tree logic* CTL. A CTL formula is composed of a path quantifier A (universal quantifier) or E (existential quantifier) followed by a linear temporal formula with temporal

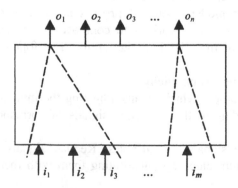

Figure 3, Irrelevant Cone Elimination

operators X, G, F, U and W. Quantifier A selects all of state paths, while quantifier E selects at least one state path. The linear temporal operator X, G, F, U and W selects the states of next clock cycles, all cycles, some cycles, until some cycles, and unless some cycles respectively. A formula is true for a description if it holds in all initial states. For instance, the CTL formula AGf represents that formula f must hold at all states reachable from the initial states in all possible paths.

4.2   Irrelevant cone elimination

The final model uses a Boolean functional vector to represent the transition functions [7]. This considerably limits the explosion of the size of the transition representation for large systems. Further more, the representation also makes it easy to eliminate parts of the model that are not relevant to the specification. The model checker needs to consider only the transition functions of the variables in the cone of influence of the specification, as figure 3 shows. In figure 3, the output o_1 of the FSM is influenced by inputs i_1, i_2, and i_3, but is irrelevant to the other inputs.

So if a property relates only with the output o_2, we can only consider the cone area (o_1, i_1, i_2, i_3) of the FSM state space and eliminate the rest area, thus reduce the number of states for model checking and speed up the process of verification further. And it's the same about the case of cone (o_n, i_m). A cone of influence is simply constructed from the true support set of the transition functions.

4.3   Property checking

Property checking is a procedure finding the set of states in the internal model where a given CTL formula is satisfied. Since the model is represented as a Boolean functional vector, the complexity of the algorithm is linear in the size of the vector and in the length of the formula. The algorithm is quite fast in practice.

First, it will traverse the model to compute the reachable states of the model, which called reachability analysis. We can use the result to further reduce the size of BDDs representing transitions.

Then it is time to check the property in the specification file by using symbolic model checking techniques. If the property is not hold, the property verifier will call the counterexample generation module [8] to produce a counterexample as a VHDL testbench. Designers can use this information to debug the design.

4.4 Counterexample generation

One of the most important features of CTL model checking algorithms is the ability to find counterexamples. When this feature is enabled and the model checker determines that a formula with a universal path quantifier is false, it will find a computation path, which demonstrates that the negation of the formula is true.

Noting that the counterexample for a universally quantified formula (prefixed by A) is the witness for the dual existentially quantified formula (prefixed by E), Our generation algorithm focus on finding witnesses for the three basic CTL operators EX, EU, and EG [8].

5. Experimental results

We have implemented VERIS in C language on SUN-SPARC workstation. The Table 1 gives some experimental results when running in the SUN-SPARC20 with 128M memory. The first column gives the circuit name. Traffic Light Controller is obtained from [9], daisy arbiter comes from [2], and priority arbiter is obtained by removing the token ring in the daisy arbiter. The second column is the model checker system we will compare each other. CV, the model checker from CMU, applies mixed abstract machine model for both synchronous and asynchronous circuits (Deharbe method) [4]. The third column shows the CPU time for modeling. The fourth column shows the CPU time to compute the set of reachable states (reachability analysis). The fifth column shows the memory usage of every example running. The sixth and seventh columns show the number of possible states and the number of reachable states from the initial states.

From the experimental results and model analysis, we can see that the VERIS requires quite less state space and memory size than the CV system. In CV, the size of state space is unreadable. Whereas in VERIS, it is agreement with our prediction. For example, 8-biter counter apparently has only 256 states, which equal the size of state space yielded in VERIS, but in CV, the size of sate space is intangible 2.15E+09.

One of the interesting examples is the 16-bit counter. Before we do irrelevant reduction, computing reachable states takes 314 seconds, and it allocated 169K memory and 1,410 BDD nodes to store the computing results. If the property to be verified just care the behavior of the lower 8 bits, it will only takes 0.36 seconds to compute reachable states, and spend 72K memory and 114 BDD nodes by using the irrelevant reduction optimization.

6. Conclusion

In this paper, we introduced a model checker for synchronous VHDL designs. The model checker applies the feature of synchronous design and the locality feature of property to reduce the state space of the internal model, thus speeding the reachability analysis and property checking of real circuits. And a counterexample generation module will invoke while some property is not hold.

In the future, we will focus on the further performance improvement. Also we will extend the VHDL subset to increase its usability.

Reference

1. J. R. Burch, E. M. Clarke, and D. E. Long, "Symbolic model checking with partitioned transition relations", In Proc. International Conference of VLSI, Aug. 1991.

2. K. L. McMillan, "The SMV System", Technical Report, School of Computer Science, CMU, 1992.

3. C. Delgado Kloos and P. Breuer, editors. Formal Semantics for VHDL, volume 307 of Series in Engineering and Computer Science. Kluwer Academic Publishers, 1995.

4. D. Deharbe and D. Borrione. Semantics of a verification-oriented subset of VHDL. In P.E. Camurati and H. Eveking, editors, CHARME'95: Correct Hardware Design and Verification Methods, volume 987 of Lecture Notes in Computer Science, Frankfurt, Germany, October 1995. Springer Verlag.

5. IEEE, "IEEE Standard VHDL Language Reference Manual", Std 1076-1993, 1993.

6. J. R. Burch, E. M. Clarke, D. E. Long, et al, "Symbolic model checking for sequential circuit verification", in IEEE Trans. On Computer-Aided Design of Integrated Circuits and Systems, Vol. 13, NO. 4, pp. 401-424, April 1994.

7. O. Caudate, C. Berthed, and J. C. Madder, "Verification of sequential machines using functional vectors", In International Workshop on Applied Formal Methods for Correct VLSI Design, volume VLSI Design Methods-II, pages 179--196, 1990.

8. E. M. Clarke, O. Grumberg, K. L. McMillan, et al, "Efficient Generation of Counterexamples and Witnesses in Symbolic Model Checking", 32nd Design Automatic Conference, 1995.

9. R. Lipsett, C. F. Schaefer, C. Ussery, "VHDL: Hardware Description and Design", Kluwer Academic Publishers, 1989.

106

10. Jinsong Bei, Jinian Bian, Hongxi Xue et al. "FSM Modeling of Synchronous VHDL Design for Symbolic Model Checking", In Proceedings of ASP-DAC'99, Hong Kong, 1999, 363-366.

11. R.P.Kurshan, Bell Laboratories, Murray Hill, "Formal Verification In a Commercial Setting", In Proceedings of 34th ACM/IEEE Design Automation Conference, 1997.

Yiping Fan is a graduate student of Department of Computer Science and Technology, Tsinghua University, Beijing, China. He received B.S. degree from Tsinghua University in 1998 and is reading for Ph.D. as a graduate student. He is researching in the technology of formal verification, especially model checking.

Jinsong Bei graduated from Department of Computer Science and Technology, Tsinghua University, Beijing, China in 1995. Then he got Ph.D. from Tsinghua University in 1999. In course of his graduate study, his research area is formal verification of model checking and BDDs. Now he is an IT practitioner.

Jinian Bian is a professor of Department of Computer Science and Technology, Tsinghua University, Beijing, China. He has been a teacher of the department since he graduated form Tsinghua University in 1970. His research area is electronic design automation (EDA), including logic and high-level synthesis, simulation and formal verification. His current interests also include design environment of System-On-a-Chip (SOC).

Hongxi Xue is a professor of Department of Computer Science and Technology, Tsinghua University, Beijing China. He graduated form Tsinghua University in 1962 and worked here as a teacher until now. His research area is electronic design automation (EDA), including synthesis, simulation and formal verification. Now he interests to developing a design environment of System-On-a-Chip (SOC).

Xianlong Hong graduated from Tsinghua University, Beijing, China in 1964. Since 1988, he has been a professor in the Department of Computer Science & Technology, Tsinghua University. His research interests include VLSI layout algorithms and DA systems. He is the Senior Member of IEEE and Chinese Electronics Association.

Jun Gu received B.S. degree from Chinese Science & Technology University in 1982 and Ph.D. degree from Utah University in 1989. He was the professor of Calgary University in Canada and currently he is the professor of the Department of Computer Science in The Hong Kong University of Science & technology. His research interests include the optimization algorithms, local search and global optimization, and their application in VLSI CAD, system engineering, communication and multi-media fields. He is the chief scientist of the 973 Key Foundation Research & Development Project "The Application Theories and High Performance Software in Information Technology" in China.

Circuit Example	Model Checker	Time of Modeling(s)	Reachability analysis(s)	Memory Usage (KB)	State space	Reachable states
Traffic Light Controller	VERIS	0.04	0.05	117	$8,192\ (2^{13})$	37
	CV	0.06	0.19	263	2.62E+05	241
8-bit counter	VERIS	0.03	0.36	230	$256\ (2^{8})$	256
	CV	0.91	18.50	368	2.15E+09	2,241
16-bit counter	VERIS	0.07	314.99	3,383	$65,536\ (2^{16})$	65,536
	CV	3.54	6,080.90	86,265	9.22E+18	5.90E+05
4-cell daisy arbiter	VERIS	0.07	0.31	307	$65,536\ (2^{16})$	1,536
	CV	0.09	2,040.70	4,374	2.15E+09	4.18E+06
16-cell daisy arbiter	VERIS	2.14	22,870.00	454	$1.85E+19\ (2^{64})$	8.03E+10
	CV	7.32	-	-	-	-
8-cell priority arbiter	VERIS	0.05	0.04	93	$65,536\ (2^{16})$	2,304
	CV	0.13	8,812.20	558	8.59E+09	5.10E+08
64-cell priority arbiter	VERIS	1.62	17.51	488	$3.40E+38\ (2^{128})$	1.20E+21
	CV	381.50	-	-	-	-

Table 1, Experimental Results

- Means data not available

SYNTHESIS

Title On Flip-flop Inference in HDL Synthesis
Subtitle

AUTHOR Hen-Ming Lin and Jing-Yang Jou
Affiliation Department of Electronics Engineering, National Chiao Tung University, Hsinchu, Taiwan 300, R.O.C.

Key words: Flip-flop inference, Multiple-clocked flip-flop, HDL (Hardware Description Language) synthesis, RTL (Register Transfer Level) synthesis, Retiming, Computer-aided design.

Abstract: In HDL synthesis at register transfer level (RTL), edge-triggered flip-flops are inferred to keep the consistence of the memory semantics between the target synthesized netlist and the original design written in hardware description language (HDL). Since typical synthesizers use ad hoc method to solve the flip-flop inference problem, either superfluous flip-flops or unreasonable limitations on coding style are necessary. Even worse, the ad hoc algorithms adopted by the typical synthesizers could incur the mismatches between synthesis and simulation. In this paper, we propose a uniform framework based on a concept called *MC flip-flop* to infer flip-flops systematically and correctly. Our approach does not impose limitations on coding style and does not infer superfluous flip-flops. Furthermore, it does not suffer from the mismatches between synthesis and simulation and can synthesize the HDL descriptions that cannot be synthesized by typical synthesizers.

1.   INTRODUCTION

The HDL synthesis is a process that converts a behavioral HDL description into an optimized structural netlist while the optimized structural netlist still keeps the semantics of the input HDL description. Therefore, two main procedures, *domain translation* and *optimization*, are involved in most synthesis tools [1], [4]. The *domain translation* translates the input HDL description from behavioral domain to structural domain and generates an

111

P.J. Ashenden et al. (eds.), System-on-Chip Methodologies & Design Languages, 111–122.
© 2001 *Kluwer Academic Publishers. Printed in the Netherlands.*

initial structural netlist that preserves the semantics of the input HDL description. The *optimization* optimizes the initial structural netlist according to the design criterions such as performance, power, testability, etc.

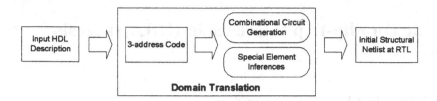

Fig. 1. The major tasks of the domain translation in most RTL synthesizers.

Fig. 1 is the flow chart of domain translation in most RTL synthesis tools. In Fig. 1, a HDL description is first translated to an intermediate form called *3-address code* [10]. Then the *combinational circuit generation* infers necessary combinational circuits for synthesizing the input HDL. The *special element inferences* including three major tasks, flip-flop inference, latch inference, and tri-state buffer inference, are to infer enough special elements (edge-triggered flip-flops, level sensitive latches, or tri-state buffers) to keep the consistencies of special semantics between the synthesized netlist and the input HDL description. In the reference [6], a novel synthesis flow is proposed to conduct those special element inferences in HDL synthesis correctly and efficiently. The related literature for the latch inference and tri-state buffer inference can be found in the references [7], [8]. In this paper, we only focus on the problem of flip-flop inference.

Since most of synthesis tools use ad hoc methods to solve this problem, designers must follow the limited coding style in their HDL descriptions in order to get the efficient and correct results. Otherwise, synthesizers may allocate superfluous flip-flops in the synthesized netlist or generate a wrong netlist whose behavior is inconsistent with the input HDL description.

In this paper, we proposed a synthesis model based on a concept called *multiple clocked flip-flops* (MC flip-flops). According to the model, we present a uniform framework based on the retiming technique [11]-[13] to infer flip-flops systematically and correctly. The result comparisons show that our approach can infer the minimum number of flip-flops systematically and correctly without incurring the unreasonable limitations on coding style.

The remainder of this paper is organized as follows. Section 2 states the preliminaries and motivations. In section 3, we propose a novel model to solve the flip-flop inference problem systematically. Section 4 discusses some synthesis considerations in the framework. The result comparisons are presented in section 5. Finally, section 6 concludes the paper.

2. PRELIMINARIES AND MOTIVATIONS

Definition 1 : In Verilog, an *edge-triggered event* is either a posedge or a negedge clause, e.g., 'posedge clock.' A *clocked statement* is an *always statement* that only includes *edge-triggered events* in its sensitivity list. The variables appearing in the edge-triggered events are called *clocks*. The *clocked statement* with only one clock is called *simple clocked statement*. Otherwise, it is a *complex clocked statement*.

Flip-flop inference must be conducted when synthesizing the clocked statements.

For simple clocked statements, typical synthesizers infer a D flip-flop for each variable being assigned in the simple clocked statements [1], [5]. For illustration in Fig. 2, typical synthesizers will infer two flip-flops as shown in Fig. 2(c) because the variables st and Y are assigned in Fig. 2(a). However, the simple clocked statement in Fig. 2(a) describes a finite state machine (FSM). The state transition graph (STG) of this FSM is shown in Fig. 2(b). Because the output Y only depends on the *state variable* st, the FSM is a *Moore machine*. Therefore, only one flip-flop corresponding to the variable st is needed for the FSM to keep the state information theoretically and the flip-flop for the variable Y is redundant.

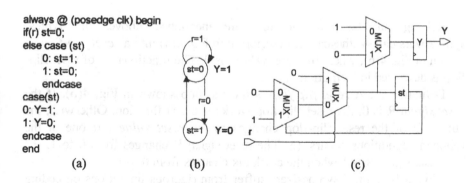

(a) (b) (c)

Fig. 2. (a) A Moore machine described in Verilog. (b) The state transition graph. (c) The initial structural netlist generated by typical synthesis tools.

For complex clocked statements, most typical synthesizers require the following assumption [1], [5].

Assumption : The variables v_i appearing in the sensitivity list except the clock of flip-flops must be used in an if-statement directly following the always statement as shown in Fig. 3. The assignments in the blocks B1 and B2 must be constant assignments. If the variable v_i is in a posedge clause in the sensitivity list, it must be one of the selectors of the if-statements.

114

Otherwise, its complemented signal '!v_i' must be one of the selectors of the if-statements.

```
always @ (posedge clock or posedge reset or negedge set)
if(! set) begin B1 end
else if (reset) begin B2 end
else begin B3 end
```

Fig. 3. The syntactic template required for typical synthesizers to synthesize complex clocked statements.

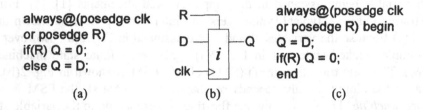

```
always@(posedge clk
or posedge R)
if(R) Q = 0;
else Q = D;
```
(a)

(b)

```
always@(posedge clk
or posedge R) begin
Q = D;
if(R) Q = 0;
end
```
(c)

Fig. 4. (a) A Verilog description for a reset flip-flop. (b) The reset flip-flop. (c) Another Verilog description for the reset flip-flop.

Relying on the syntactic template mentioned above, the typical synthesizers can synthesize the description in Fig. 4(a) into a *reset flip-flop* as shown in Fig. 4(b), where the reset value *i* is 0. The functionality of reset flip-flop is described in Definition 2.

Definition 2 : A *reset flip-flop* is the flip-flop shown in Fig. 4(b). If the *reset signal* R is 0, the reset flip-flop works like a D flip-flop. Otherwise, the output Q of the reset flip-flop changes to the *reset value i* if one of the following conditions occurs. (a) The reset signal R changes from 0 to 1. (b) The reset signal R is 1 when the clock clk changes from 0 to 1.

Obviously, typical synthesizers suffer from rigorous limitations on coding style. For example, the behavior of the complex clocked statement shown in Fig. 4(c) is the same as that of the one shown in Fig. 4(a). However, typical synthesizers cannot synthesize the description in Fig. 4(c) because the description in Fig. 4(c) does not satisfy the syntactic template described in Fig. 3, i.e., the if-statement does not follow the always statement immediately.

3. OUR APPROACH

Our synthesis model for both simple and complex clocked statements is shown in Fig. 5(a). It consists of a data path and a 'fence' being controlled by the clocks in the clocked statement. This synthesis model comes from the following observations. A clocked statement can be conceptually divided into two parts: (1) a set of statements that describes how the data manipulation is performed in the clocked statement; (2) a set of clocks that prevents the results of the data manipulation from being assigned to outputs until one of the clocks is triggered. The data path and the fence in our synthesis model implement these two parts, respectively.

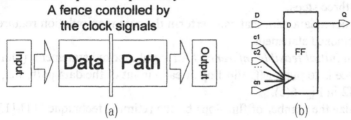

(a)   (b)

Fig. 5. (a) Our synthesis model. (b) The MC flip-flop.

The flip-flop inference problem is to implement the 'fence' with the minimum number of the flip-flops being controlled by the clocks. To implement the fence in our synthesis model, the MC flip-flop described in Definition 3 is useful.

Definition 3 : The *multiple clocked flip-flop*, *MC flip-flop* for short, is the flip-flop with a set of clocks, c1, c2, ..., cn, as shown in Fig. 5(b). The output value Q of MC flip-flop changes to the input value D if one of the clocks changes from *0* to *1*. Otherwise, the output Q keeps its previous value.

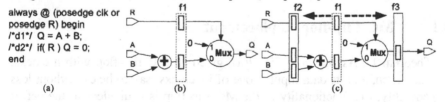

(a)   (b)   (c)

Fig. 6. (a) A clocked statement. (b) The fence in our synthesis model. (c) The motion of the fence in our synthesis model.

When synthesizing a clocked statement with clock set C, the fence in our synthesis model can be implemented by inserting exact one MC flip-flop with the clock set C into each path that starts from the input of the data path and ends at the output of the data path. For example, the data path in Fig. 6(b) can perform the data manipulation described by the statements d1 and d2 in Fig. 6(a). The fence f1 in Fig. 6(b) that consists of two MC flip-flops with clocks clk and R is a feasible inference because it results in that the value changes on inputs R, A and B can only affect the output Q after one of the clocks is triggered. Furthermore, we can move the fence forward and backward as shown in Fig. 6(c). Obviously, both fences f2 and f3 are feasible while the number of MC flip-flops needed for constructing the fences is different. The motion of fence is similar to the *retiming technique* in sequential logic minimization [11]. Therefore, the Integer Linear Programming (ILP) formulation proposed in the reference [11] can be used to find the minimum flip-flop inference and our approach consists of the following three steps.

1. Construct a data path that can perform the data manipulation required by input clocked statement.
2. Find an *initial feasible inference I* for input clocked statement. (Our initial inference is to put a MC flip-flop at each input of the data path, i.e., the fence f2 in Fig. 6(c).
3. Minimize the number of flip-flops by the retiming technique [11]-[13].

4.   COPE WITH COMPLEX CLOCKED STATEMENTS

Given a clocked statement with n clocks, if n = 1, the MC flip-flops used in our approach can be implemented by D flip-flop. However, if n > 1, generally it has no the MC flip-flop with n clocks in the synthesis library. To conquer this issue, we need to eliminate the MC flip-flops with multiple clocks in our initial inference. The observations and Theorems described in the following subsections 4.1 and 4.2 are useful for this purpose.

4.1   MC Flip-flop Replacement

Theorem 1 : As shown in Fig. 7(a), given an MC flip-flop with n clocks, $c1, c2, ..., cn$, and its data input is one of its clocks, say to be $c1$ without loss of generality, the functionality of the MC flip-flop is equivalent to the netlist shown in Fig. 7(b) that contains an AND gate and n-1 reset flip-flops whose reset signals are all $c1$, reset values are all 1, data inputs are all 0, but clocks are $c2, c3, ..., cn$, respectively.

Proof: First, Fig. 7(c) shows the HDL model of the MC flip-flop in Fig. 7(a). In Fig. 7(b), all flip-flops are the reset flip-flop mentioned in Definition 2 and their reset signals are all $c1$. Furthermore, each flip-flop only has one clock, which is $c2, c3, ..., cn$, respectively. Therefore, if $c1$ is triggered or the value of $c1$ is *1* and the other clocks are triggered, all outputs of the reset flip-flops are *1* due to the reset value of flip-flops. It results in that the output of the AND gate is *1*. If the value of $c1$ is *0* and a clock ci is triggered, the output of the flip-flop with clock ci is *0*, where i could be 2, 3, ..., n. It results in that the output of the AND gate is *0*. The behaviors described above are exactly the behavior of the HDL model shown in Fig. 7(c) because the assignment 'Q = c1' is executed after one of the clocks already changes from 0 to 1.

Please note that after replacing the MC flop-flop in Fig. 7(a) by the netlist shown in Fig. 7(b), no MC flip-flop with multiple clocks is needed.

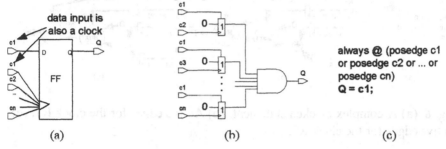

(a)   (b)   (c)

Fig. 7. (a) An MC flip-flop with its data input being one of its clocks. (b) The equivalent implementation using reset flip-flops and AND gate. (c) The HDL model.

4.2   Redundant Clock Elimination

Definition 4 : An edge (u, v) is *passive* for a clock c, if the output value of the vertex u cannot affect the output of the vertex v from the moment that the clock c changes from 0 to 1 to the moment that another clock is triggered. Otherwise, the edge (u, v) is *active* for the clock c. A path $(u_0, u_1, ..., u_k)$ is *active* for a clock c if each edge (u_i, u_{i+1}), where i = 0, 1, ..., k-1, is active for the clock c. Otherwise, it is *passive*.

For example, Fig. 8(b) shows our initial inference for the complex clocked statement shown in Fig. 8(a). In Fig. 8(b), the bolded edges are active and the thin edges are passive for the clock R1, respectively. The reason is as follows. When R1 changes from 0 to 1, the selective signal of Mux1 should be set to 1 according to the HDL description in Fig. 8(a) and it results in that the output of Mux1 being set to 0. The edges (A, +), (B, +), (R1, Mux1) and (0, Mux1) are active for the clock R1 because when R1 changes from 0 to 1, the vertices A and B can affect the output of adder and the vertices R1 and the 0 can affect the output of Mux1. However, the edge (+, Mux1) is passive because the output of adder cannot affect the output of Mux1. At the same time, because both two data inputs of Mux2 are set to 0, the selective signal of Mux2 cannot affect the output of Mux2. Therefore, the edge (R2, Mux2) is passive and the edges (0, Mux2) and (Mux1, Mux2) are active. The bolded edges in Fig. 8(c) further show the active edges for the clock R2.

Furthermore, the path <R2, Mux2, Q> is passive for the clock R1. Therefore, the output of the MC flip-flop on the edge (R2, Mux2) cannot affect the output Q from the moment that the clock R1 is triggered to the moment that another clock is triggered.

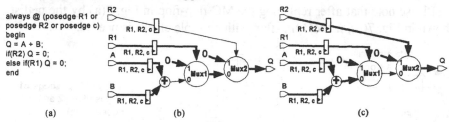

Fig. 8. (a) A complex clocked statement. (b) Active edges for the clock R1. (c) Active edges for the clock R2.

Note that if a clock c triggers a MC flip-flop F to sample a value F_c and F_c cannot affect the output of data path of the input clocked statement, the clock c is redundant for the MC flip-flop F. The reason is that all MC flip-flops in our initial inference have the same clock set and they will overwrite their value sampled when the clock c is triggered if there is any clock being triggered again. Without loss of generality, assume that each clock c_i is triggered when the value of c_i changes from 0 to 1. Using the path activation, the following steps can eliminate the redundant clocks for the MC flip-flops in our initial inference.

For each clock c_i which is one of the inputs of data path,
1. Set c_i to 1. (to 0 if c_i is triggered when its value changes from 1 to 0.)
2. Compute the deterministic values caused by the step 1.
3. For each edge (u, v) in the data path, determine whether the edge (u, v) is active for the clock c_i.
4. For each input u_l, if each path that starts at u_l and ends at the output of the data path is passive, eliminate the clock c_i for the MC flip-flop being located at the input u_l.

In Fig. 8(b), each path starting from the vertices R1, A, or B, and ending at the vertex Q is passive for the clock R1. Therefore, we can eliminate the clock R1 for the MC flip-flops without being shaded in Fig. 8(b). Applying the same way, we can eliminate the clock R2 for the MC flip-flops without being shaded in Fig. 8(c).

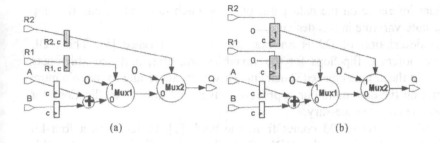

Fig. 9. (a) Our initial inference after eliminating redundant clocks. (b) Our initial inference for further minimization.

Fig. 9(a) further shows the state of the MC flip-flops in our initial inference after eliminating the redundant clocks. As describing in section 4.1, the top two MC flip-flop in Fig. 9(a) can be further replaced by reset flip-flops because their data inputs are also one of their clocks and Fig. 21(b) shows our initial inference for further minimization. The minimization can be conducted by using the *bound retiming* that was proposed in the reference [13] and can correctly retime the reset flip-flops. In this example, we can eliminate all MC flip-flops with multiple clocks. Please note that after conducting the MC flip-flop replacement and redundant clock elimination, if there are still MC flip-flops with multiple clocks in our initial inference, we report that the input complex clocked statement cannot be synthesized.

5.   RESULT COMPARISONS

We compare our results to those produced by the *HDL Compiler* of *Synopsys®*. Fig. 10 and Fig. 11 emphasize the comparisons to simple clocked statements and complex clocked statements, respectively. In Fig. 10, there are three designs D1, D2 and D3. Their descriptions and HDL Compiler's results are shown in column 1 and column 2. Column 3 shows the data path constructed by our approach. In column 3, please note that the minimum inference is shown by the shaded boxes that make each path starting from the input of data path and ending at the output of data path have one flip-flop. Column 4 shows the netlists generated by our approach.

The description of D1 represents a Moore machine whose output Y only depends on the state variable. Therefore, only one flip-flop for the state variable is necessary. However, using the limited algorithm described in section 2, the HDL Compiler infers 2 flip-flops for this Moore machine, one for the state variable st and the other one for the output Y. According to the minimum inference on the data path, our approach only infers one flip-flop for the state variable in this design.

The description of D2 is an example in the reference [5]. The HDL Compiler infers 6 flip-flops for the variables cnt, abit, obit and xbit being assigned in the simple clocked statement. However, according to the minimum inference on the data path, our approach derives that only three flip-flops for the variable cnt are necessary.

The description of D3 comes from the book [1]. It describes a four-bit shift register. However, the HDL Compiler infers 7 flip-flops for this description, 3 for the variable pipe and 4 for the variable Y. In this example, our approach can derive that only 4 flip-flops are necessary.

	Verilog descriptions	Synopsys results	Data graph			Our results
D1	always @ (posedge clk) begin if(R) st=0; else case (st) 0: st=1; 1: st=0; endcase case(st) 0: Y=1; 1: Y=0; endcase end		V^S	V^F	V^D	
D2	input clk, R; output abit, obit, xbit; reg abit, obit, xbit; reg [2:0] cnt; always @ (posedge clk) begin if(R) cnt = 0; else cnt = cnt +1; abit = & cnt; obit =	cnt; xbit =^ cnt; end		V^S	V^F	V^D	
D3	input D; output Y; reg [3:0] pipe, Y; always @ (posedge clk) begin for(n=3; n>=1; n=n-1) pipe[n]=pipe[n-1]; pipe[0]=D; Y=pipe; end		V^S	V^F	V^D	

Fig. 10. Result comparisons on simple clocked statements.

In Fig. 11, there are two designs D4 and D5. Their descriptions, HDL Compiler's results, our initial inference and our results are shown in column 1, 2, 3 and 4, respectively.

The description of D4 consists of a complex clocked statement. Our approach can synthesize a reset flip-flop with reset value 0 and reset signal being R1+R2 for the description D4. However, HDL Compiler cannot synthesize this design and reports an error, because this description does not satisfy the syntactic templates provided by HDL Compiler.

For the description D5, HDL Compiler infers a flop-flop with asynchronous set and reset for this design and our approach derives that 3 reset flip-flops are necessary. The most important thing is that the result of HDL Compiler is a wrong inference and it suffers from the mismatch between synthesis and simulation. This is demonstrated in Fig. 12. Fig. 12 shows the simulation wave forms of the original description, HDL Compiler's result and the result of our approach. We can see that the wave form of HDL Compiler's result is not consistent with the original one in the circle of Fig. 12. At the moment, the reset signal changes from *1* to *0* and this event cannot trigger the execution of the complex clocked statement in D5. However, the inference of HDL Compiler re-computes the value of output *Q*. This mismatch could results in extra overhead on verifying a design. Please note that our result does not incur this mismatch in Fig. 12.

	Verilog descriptions	Synopsys results	Our initial inference	Our results
D3	always @ (posedge R1 or posedge R2 posedge c) begin Q = A + B; if(R2) Q = 0; else if(R1) Q = 0; end	Not Available !!		
D4	always @ (posedge R or posedge S or posedge c) if(R) Q = 0; else if(S) Q = 1; else Q = D;			

Fig. 11. Result comparisons on complex clocked statements.

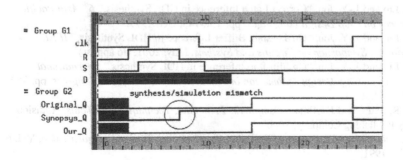

Fig. 12. The simulation waveforms of the design D5.

122

6. CONCLUSIONS

Since typical synthesizers use ad hoc method to solve the flip-flop inference problem, either superfluous flip-flops or unreasonable limitations on coding style are necessary for synthesizing a clocked statement. Even worse, the ad hoc algorithms could incur the mismatches between synthesis and simulation.

Comparing to typical ad hoc approaches, our approach has no limitation on coding style. For simple clocked statements, our approach does not infer superfluous flip-flops. For complex clocked statements, our approach can synthesize the designs that cannot be synthesized by typical synthesizers without limitations on coding style. The most important thing is that our approach does not suffer from the mismatches between synthesis and simulation.

REFERENCES

[1] D. J. Smith, *HDL Chip Design: A Practical Guide for Designing, Synthesizing and Simulating ASICs and FPGAs Using VHDL or Verilog.* Doone Publications.
[2] IEEE, New York. *VHDL Language Reference Manual,* IEEE Standard 1076-1993, June 1994.
[3] *Verilog-XL Reference Manual,* Cadence Design Systems Inc., 1991.
[4] *Design Compiler Family Reference Manual,* Synopsys Inc., 1994.
[5] *HDL Compiler for Verilog Reference Manual Version 3.1a,* Synopsys Inc., Mar. 1994.
[6] H.-M. Lin, *On HDL Synthesis at Register Transfer Level and Related Graph Theory,* Ph. D. Thesis, Chiao Tung University, Taiwan, 2000.
[7] H.-M. Lin and J.-Y. Jou, "Correct Latch Inference in HDL Synthesis," *6th Asia Pacific Conference on Chip Design Languages,* pp. 138-143, Oct. 1999.
[8] H.-M. Lin and J.-Y. Jou, "On Tri-state Buffer Inference in HDL Synthesis," *IEEE International Symposium on Circuits and Systems,* May. 2001. (to appear)
[9] H.-M. Lin and J.-Y. Jou, "On Flip-flop Inference in HDL Synthesis," *International Conference on Chip Design Automation in 16th IFIP World Computer Congress,* pp. 369-376, Aug. 2000.
[10] Aho, R. Sethi, and J. Ullman, *Compilers: Principles, Techniques and Tools.* Addision-Wesley Publishing Company, 1986.
[11] C. E. Leiserson and J. B. Saxe, "Retiming Synchronous Circuitry," *Algorithmica,* Vol. 6, pp. 5-35, 1991.
[12] V. Singhal, S. Malik and R. K. Brayton, "The Case for Retiming with Explicit Reset Circuitry," *International Conference on Computer-Aided Design,* pp. 618-625, 1996.
[13] K. Eckl, J. C. Madre, P. Zepter and C. Legl, "A Practical Approach to Multiple-Class retiming," *36th Design Automation Conference,* pp. 237-242, 1999.

Synthesis Oriented Communication Design for Structural Hardware Objects

Wolfram Putzke-Röming, Wolfgang Nebel
OFFIS Research Center, Escherweg 2, 26121 Oldenburg, Germany

Abstract   An object-oriented HDL has to provide an appropriate communication concept
for communication between objects. Unfortunately, exactly the communication
concept can be identified often as a critical issue in several existing proposals for
object-oriented HDLs. In this paper we present requirements for such a
communication concept and exemplary define a new object- and synthesis-
oriented communication model for Objective VHDL.

1.   Introduction

In recent years several proposals have been made for object-oriented extensions of VHDL. The analysis of these proposals shows that there are basically two approaches to integrate object-orientation into VHDL. One
approach, which is inspired by object-oriented programming languages, is the
object-oriented extension of the VHDL type system. The other approach is the
object-oriented extension of the VHDL design entity. In this paper only the latter approach is considered which provides some clear advantages:
• A design entity can be active by itself.
• A design entity is a very intuitive unit for object-oriented classification.
• Structural partitioning is possible at object-oriented design level.
However, in spite of these advantages most of the entity based approaches are
lacking an appropriate communication mechanism for the structural hardware
objects. This is a serious problem for these approaches since communication
between objects is indispensable for object-oriented design.

To solve this problem the first question to answer is: What are the requirements for a communication mechanism? The object-oriented VISTA approach
[16] for example provides a very powerful and abstract communication concept. However, the cost of this concept is a missing link for communication
synthesis. On the other hand Objective VHDL [8] leaves it entirely to the user
to specify all details of communication between structural objects. This is also
unacceptable due to the high modeling cost for communication.

Furthermore, there are several works outside the domain of object-oriented
extensions to HDLs which seems to be valuable for definition of an appropriate communication concept. Two directions should be mentioned here: First,
there is some work on the separation of component behavior and communication design [15]. Approaches which follow this idea are VHDL+ [7] and

123

P.J. Ashenden et al. (eds.), System-on-Chip Methodologies & Design Languages, 123–134.
© 2001 *Kluwer Academic Publishers. Printed in the Netherlands.*

SpecC [5]. Especially for VHDL+ a significant increase of productivity is reported. Secondly, there is some fundamental work on object-oriented communication design and refinement [4][6][18]. However, most of this work is done for the software domain and can not be transferred directly to hardware (cf. section 3.2).

The structure of this paper is the following: After a brief introduction of Objective VHDL some requirements for communication between structural objects are identified. Based on these requirements a new communication model is proposed and integrated into Objective VHDL by some language extensions. Due to the restricted space of this paper the language extensions are introduced only informally. Finally, some ideas for synthesis of the proposed communication model are presented.

2. Objective VHDL—a brief overview

Objective VHDL [8] is an object-oriented extension to VHDL which has been designed within the OMI-ESPRIT project REQUEST. Currently, several tools for the application of Objective VHDL are available[1].

Objective VHDL, in difference to most competitive approaches, supports object-oriented extensions based on the VHDL type system (class types) as well as structural objects based on the VHDL design entity (entity objects). While application and benefits of the class types have been discussed in several publications [13] the entity objects are used very rarely. The main reason for this is that there is no appropriate communication mechanism for entity object communication.

2.1. Entity Objects

The Objective VHDL entity objects are based on the intuitive idea that a structural hardware component represented by a VHDL design entity can be considered as an object in the object-oriented meaning. Similar to an object-oriented object the design entity allows to specify data which defines the state of the entity and allows one to implement some functionality. Finally, the design entity has a well defined interface for communication. Due to this affinity of the design entities with the idea of objects, VHDL can be classified as an object-based modeling language [9].

Example 1 shows the object-oriented interpretation of an Objective VHDL entity object. The syntax of an Objective VHDL entity object remains widely unchanged compared to the VHDL design entity. However, the interpretation of parts of the design entity has changed.

---

1. http://eis.informatik.uni-oldenburg.de/research/objective_vhdl.shtml

Shared variables or signals declared within an entity object will be considered as class attributes in the object-oriented meaning while procedure or function declarations will be considered as class methods.

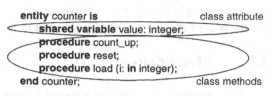

Example 1. Entity Object counter

For specialization of entity objects an inheritance mechanism for the entity objects is provided. For detailed syntax and semantics of the entity object inheritance please refer to the Objective VHDL LRM [8]. Example 2 shows an derived entity object.

As usual with inheritance a derived entity object can add or redefine class attributes and methods. In the example the method count_down is added while the class attribute and the methods

```
entity up_down_counter is new counter with
    procedure count_down;
end up_down_counter;
```

Example 2. Derived entity object

from the parent entity object counter are inherited.

However, special emphasis has to be given to the communication interface of a VHDL design entity and an entity object respectively. Both are basically the same, due to the fact that an entity object is only an object-oriented representation of a VHDL design entity. Hence, the communication interface is characterized by the instantiation name of the entity (object), the port list, and the generic list.

Unfortunately the usage of the communication interface of the VHDL design entity for the entity objects causes some problems because the communication interface differs significantly from the interface that is typically expected for an object. This difference causes two problems:

- Since the entity objects can communicate only via their ports, object-oriented communication needs to be transformed into a port based signal communication explicitly.
- The procedures declared within a VHDL design entity are completely encapsulated by the entity. Hence, the methods of an entity object are not visible to other entity objects.

While the solution of the first problem requires a sophisticated method to map object-oriented communication which has typically the style and semantics of a remote method call to a signal communication, the latter problem requires relaxation of the encapsulation of the design entity.

3. Requirements for communication

The intention of this section is the definition of requirements which have to be considered when object-oriented communication between structural hardware objects is discussed. However, due to the restricted frame of this paper the requirements capture is done on a more or less abstract level.

3.1. Methodology

The extension of an object-oriented language with appropriate communication concepts and a methodology which defines how communication is specified or refined requires a conceptual correspondence of the extensions with the fundamental concepts of the language. We will try to specify requirements for this correspondence more precisely:

- *No break in methodology*: This requirement is twofold. (a) To avoid unnecessary complexity, specification/modeling techniques should be applicable to all parts of a design—especially the communication parts. (b) Further, it is necessary that the complete design process can be captured by one design methodology. This explicitly includes the specification and refinement of communication (cf. Section 3.2). Indeed, these requirements are strongly related to issues of language design [2] [19]. Especially (a) is based on the idea of orthogonal language design which means that the concepts of a language should be simple and orthogonal applicable. Concepts or techniques which integrate not fully orthogonal make a language difficult to understand and bear the problem of potential incompatibility of such concepts.
- *Abstraction*: The communication concepts of a language should correspond with the intended abstraction level of the language. Hence, for the communication between the Objective VHDL entity objects a communication based on remote method call semantics is required in order to avoid manual and clumsy encoding of a remote method call.
- *Reuse*: The specification of the communication code (e.g. protocols) as well as the specification of the communication components can quickly become quite complex. Hence, it is desirable that communication code as well as the specification of the communicating components are reusable. This of course requires the separate (orthogonal) specification of communication code and communicating components and the possibility to store them in libraries for reuse.
- *Synthesis*: Objective VHDL was designed with special emphasis to synthesizability of the object-oriented extensions. Also the requirement of synthesis is twofold. On the one hand the design methodology should integrate the synthesis steps, on the other hand all (potential) modifications of the language must preserve the synthesizability.

3.2. Communication refinement

In the software world the process of communication specification and refinement typically follows the model of communication layers as depicted in Figure 1. A special layer represents a certain abstraction level and provides an interface to a more abstract layer as well as to a less abstract layer. Layer n denotes a refinement of layer n+1.

Since this model is intuitive and well established, we will adapt it for communication specification and refinement in the hardware world. However, the model needs some modifications to make it applicable for hardware domain:

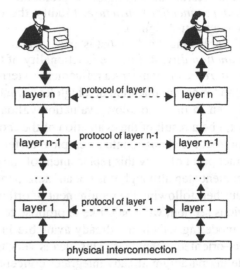

Figure 1: Model of communication layers

- In software the model often implies the semantics of a communication stack where a communication request is propagated through the layers down to the physical interconnection for real transmission. The adaptation of this semantics for hardware would mean that all communication layers must be simulated and/or synthesized in order to perform a communication. While this should be possible for very small examples, it would result in very high cost for simulation and synthesis for real world examples. Hence, we would like to restrict the model of communication layers in hardware domain only to describe the steps of communication refinement not to represent a communication stack.

- While in software the layer of physical interconnection is assumed to be existent, it is subject of communication design in hardware.

Furthermore, a more detailed consideration of typical real world examples shows that communication refinement can be subdivided into two kinds of refinement. We will call them horizontal refinement and vertical refinement. Both are illustrated in Figure 2.

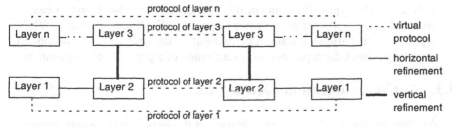

Figure 2: Horizontal vs. vertical refinement

Horizontal refinement means a real refinement of the communication protocol (e.g. addition of a control signal, refinement of transmitted data, ...). Generally horizontal refinement can be classified by four dimensions [3][17]:

- *Time*: Refine the timing of the protocol.
- *(Interconnection) structure*: Modify the (interconnection) structure (e.g. add a control signal).
- *Data*: Refine the data that is exchanged.
- *Functionality*: Refine the functionality of the protocol.

Vertical refinement means a refinement in terms of replacement. For example, assume for communication of a high level model a simple handshake protocol is chosen in order to allow evaluation/validation of the model. Later on — might it be a result of the evaluation of the model or a external requirement— the protocol is changed into a complete different one —let it be PCI. From an abstract point of view this replacement of the protocol can be considered as a refinement step although it is not an incremental refinement.

Further, following the requirement of orthogonal language design from the previous section, we would like to propose to use techniques for communication modeling which are already available in Objective VHDL —especially object-oriented techniques. This idea is very natural since the object-oriented modeling paradigm already integrates concepts for (incremental) refinement.

For the introduced kinds of communication refinement the following modeling techniques can be applied:

- *Horizontal refinement*: The basic idea to allow horizontal refinement is the application of object-oriented inheritance which is a very powerful mechanism for incremental refinement. Of course, this requires that protocols can be defined as a kind of object-oriented class which is the basis for inheritance. It has been discussed in [12] that inheritance is suited to represent the four dimensions of horizontal communication refinement which have been mentioned above.
- *Vertical refinement*: Objective VHDL provides two concepts which are suited to represent the replacement of a communication protocol as a refinement step. One concept is the object-oriented polymorphism, provided communication protocols can be modeled polymorphic. The other concept is the extension of the VHDL genericity by some kind of type genericity as available in Ada [1]. If communication protocols can be described by Objective VHDL types, the concept of type genericity allows one to pass a protocol as a generic parameter to an Objective VHDL design entity. While type genericity allows a static replacement of the protocol polymorphism generally allows the replacement or exchange of a protocol during runtime.

3.3.   Communication Mechanism

Another requirement for communication is the definition of a communication mechanism which must be compatible with the methodology for communication design. Such communication mechanism defines the technical aspects of communication as demonstrated in Figure 3.

Some of these aspects are described in the following. A detailed discussion can be found in [10][11].

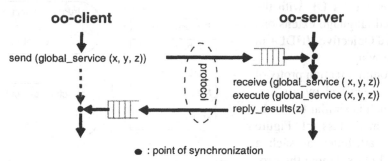

Figure 3: Aspects of a comm. mechanism

- *Addressing of objects*: While it is quite natural for a HDL that structural hardware objects can be considered to be static, it has to be decided whether the objects' interconnections should be static, too. In [16] a handle mechanism for addressing is described which allows an object to "learn" about other objects during its lifetime and to communicate to these objects.
- *Communication protocol*: As shown in Figure 3 a communication protocol is required for the exchange of information between the communication partners. While on abstract level the remote method call, which is typically used in object-oriented languages, can be considered as a kind of communication protocol, the lower more synthesis oriented levels require a more detailed communication protocol.
- *Synchronization*: Since structural objects basically are concurrent, they need to synchronize themselves. Although the preferred remote method call implies a synchronous communication, it can be realized also on an asynchronous communication mechanism as indicated in Figure 3.
- *Concurrency*: Structural objects are concurrent. However, the Objective VHDL entity objects allow a finer granular concurrency by definition of several processes which define the behavior of an entity object. The general problem of this intra-object concurrency is how to ensure the consistency of the class attributes if they are accessed concurrently. The definition of a communication mechanism can imply strong consequences for the intra-object concurrency.

Another very important question is whether the communication mechanism should be fixed or flexible. While a fixed mechanism means that all aspects of the mechanism are unchangeably predefined, a flexible mechanism allows modification —at least of some parts— of the mechanism.

4.   Communication Concept

Based on the discussions and requirements of the last sections now a communication concept for communication between structural entity objects and some language extensions to Objective VHDL are proposed. The concept is

130

based on previous work [10][11].

Objective VHDL with the proposed language extensions is named Objective VHDL+ in the following.

A communication protocol will be described by a structure which is similar to an object-oriented class (cf. Figure 4). The attributes of such a protocol-class denote the protocol signals and the methods define the protocol functionality.

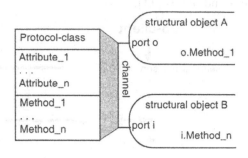

Figure 4: Oo-communication channel

Example 3 demonstrates the implementation of 3-way synchronous handshake protocol with a class-like structure. Since the semantics of the proposed protocol classes differ from the Objective VHDL type classes, a new construct —the channel type— is introduced.

```
type syn_handshake is channel
    generic ( type gen_type is new data_type);
    signal data: gen_type;
    signal data_valid: bit;
    signal ack: bit;
    signal ready_for_data: bit;
    procedure send (i: in gen_type, signal clk: in bit);
    procedure receive(o:out gen_type,signal clk:in bit);
    -- internal methods
    procedure wait_on_ready (signal clk: in bit);
    procedure wait_on_ack (signal clk: in bit);
    procedure wait_on_data_valid (signal clk: in bit);
end syn_handshake;

type syn_handshake is channel body
    ...
end channel body syn_handshake;
```

Example 3. Synchronous handshake

As described above the attributes (signals) of syn_handshake represent the protocol signals and the methods implement the protocol functionality. The interface of the protocol class is defined by its set of methods. Additionally, type genericity is introduced to Objective VHDL+. Example 3 shows the generic type gen_type which must be determined at the time of instantiation of syn_handshake.

Furthermore, communicating structural objects are interconnected by a static communication channel (cf. Figure 4) which has some similarities to a VHDL signal. The type of a channel determines the communication protocol that is used with the channel. If for example the type of a channel from Figure 4 is syn_handshake from Example 3, communication over this channel follows the synchronous handshake protocol. Similar to VHDL signals channels are mapped to entity ports which allow the identification of a channel. Hence, in Figure 4 the methods of the channel (protocol class) can be accessed via the port names.

The proposed ideas of static interconnection of communicating objects and the protocol classes are in line with the property of hardware to be static[2]. Furthermore, these ideas help to fulfill some of the requirements which are defined in the previous sections:

- *Orthogonalization (Separation)*: Example 3 shows that the communication protocol is defined by a separate class structure. Refinement of the protocol is widely independent of the communicating objects.
- *Abstraction*: The communication protocol is abstracted by the method interface of a protocol class.
- *Addressing*: The communication channels provide a mechanism for indirect addressing the communication partner by a channel.
- *Flexibility*: There is no predefined communication protocol. The protocols must be defined explicitly or can be chosen from a library. Further, the protocol can be defined independent from the data that should be transmitted due to type genericity which is introduced to Objective VHDL+.

Finally, it should be mentioned that an inheritance mechanism for channels is available similar to the inheritance of class types.

4.1. Communication model

Now the parts of the Objective VHDL+ communication model are put together to a complete model.

Object-oriented communication is characterized by a client server communication where the server provides services which can be requested by the client.

Figure 5 shows such client and server. Due to (too) strong entity object encapsulation the list of services (methods) is represented by an interface[3].

Communicating objects are interconnected by a communication channel which is an instantiation of a protocol class. The protocol class defines the communication protocol. To make the whole

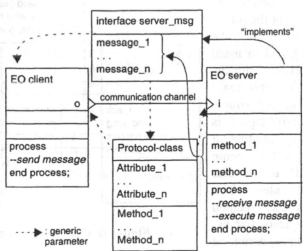

Figure 5: Objective VHDL+ comm. model

---

2. The special case of dynamically reconfigurable hardware is not considered here.
3. Interfaces are not described in this paper. In the interfaces messages are defined which correspond to the methods of the target entity object.

132

mechanism as flexible as possible the concept of type genericity is introduced.

So the entity objects and the communication protocol can be modeled widely independent of the interfaces (messages to transmit). Further, the communication protocol of a communication channel can be changed easily.

Note: Although communication channels are directed as VHDL signals, the direction reflects only the client-server relationship. A channel can contain signals in both directions.

Communication is performed by invocation of the protocol class's methods which can be accessed over the port names (e.g. o.send(...), provided send is a method of the protocol class).

5. Synthesis ideas

Since Objective VHDL+ targets hardware description, in the following the main idea for synthesis of the proposed communication mechanism is described.

Figure 6 demonstrates how the protocol classes, and that is the main new concept, can be transformed into synthesizable VHDL.

The attributes of the protocol class which represent the data and control signals of the protocol can be mapped to usual VHDL signals. While this mapping for simple control signals is very intuitive, it is a little more difficult for data attributes which represent method requests. For example, assuming

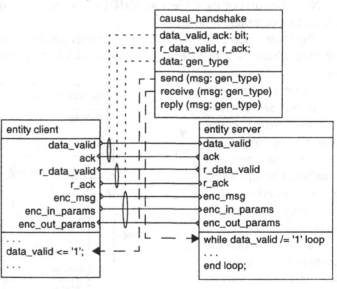

Figure 6: Communication synthesis

a client entity object communicates with the counter entity object using the causal_handshake. The messages which can be sent to the counter are defined by the counter_interface. Hence, the attribute data of causal_handshake denotes such a message.

To map such a complex attribute to usual VHDL signals the following steps are necessary. Based on the interface definition an encoding for all messages (names) can be found. This might be a binary encoding which can be used for transmission. If the associated operation of a message has parameters, they have to be transmitted, too. To do so, the parameters could be transmitted by additional signals. Figure 6 shows that the attribute data of causal_handshake can be transformed into three kinds of signals. One signal to encode the message name (enc_msg), a set of signals for the parameters of mode in (enc_in_params), and a set of signals for the parameters of mode out (enc_out_params).

Another, maybe more efficient method to map the data attributes of the protocol classes to VHDL signals would be to find one encoding for the complete messages which avoids the additional signals for in and out parameters. This method is described in [14].

The methods of the protocol classes can be inlined in the communicating objects for synthesis. However, this requires that the methods are modeled in a synthesizable style.

To automate the communication synthesis, a tool could be implemented which transforms the proposed extensions for communication into synthesizable VHDL. As described above such tool has to encode the messages, to generate VHDL signals to transmit encoded messages, and to inline protocol functionality into communicating objects.

6.   Conclusion

After a brief introduction of the Objective VHDL entity objects and the observation that Objective VHDL lacks an appropriate communication mechanism for communication between entity objects some problems have been identified which make the definition of such a mechanism rather difficult.

In order to allow the definition of an appropriate communication model several requirements have been identified which should be covered by the communication model. These requirements have been classified into requirements with methodological background and requirements based on the technical realization of communication. Based on the requirements a new concept for communication between structural objects has been proposed. To allow a smooth integration of this concept into Objective VHDL some language extensions (Objective VHDL+) have been defined. Finally, the general ideas for synthesis of the proposed communication model have been proposed.

For evaluation of the proposed communication model several communication protocols have been implemented, refined, simulated, and partly synthesized. Since the available Objective VHDL tools currently do not support the proposed language extensions, the protocol implementations have been transformed into standard (Objective) VHDL according to the synthesis ideas presented in Section 5.

134

7. References

[1] ANSI/MIL-STD-1815A-1983: *The Programming Language Ada, Reference Manual*, Springer-Verlag 1983

[2] P. J. Ashenden, P. A. Wilsey: *Principles for Language Extension to VHDL to Support High-Level Modeling*. Technical Report TR-03/97, Department of Computer Science, University of Adalaide, Australia; http://www.cs.adelaide.edu.au/users/petera/suave.html

[3] W. Ecker: *Neue Verfahren für den Entwurf digitaler Systeme mit Hardwarebeschreibungssprachen*, Dissertation TH. Darmstadt, Shaker Verlag 1996

[4] J. Ellsberger, D. Hogrefe, Amardeo Sarma: *SDL: Formal Object-oriented Language for Communicating Systems*, Prentice Hall Europe 1997

[5] D. D. Gajski, R. Dömer, J. Zhu: *IP-Centric Methodology and Design with the SpecC Language*, System-Level Synthesis, Kluwer Academic Pubishers, 1999

[6] K. Granoe, J. Harju, T. Jaervinen, T. Larikka, J. Paak: *Object-Oriented Protocol Design and Reuse in Kannel*, Design of Hardware/Software Systems, September 1995, Como, Italy - Proceedings of EUROMICRO Conference 1995

[7] ICL: *VHDL+ Language Reference Manual*, Version 5.0, http://www.icl.com/da/vhdl+.htm

[8] S. Maginot, W. Nebel, W. Putzke-Röming, M. Radetzki: *Final Objective VHDL language definition*. REQUEST Deliverable 2.1.A (public), 1997. http://eis.informatik.uni-oldenburg.de/research/request.html

[9] D. Perry: *Applying Object Oriented Techniques to VHDL*. Proceedings of the VIUF Spring Conference, 1992

[10] W. Putzke-Röming, M. Radetzki, W. Nebel: *A flexible Message Passing Mechanism for Objective VHDL*. Proceedings of the DATE 1998 Conference, pp. 242-249.

[11] W. Putzke-Röming, M. Radetzki, W. Nebel: *Modeling Communication with Objective VHDL*, Proceedings of Spring IVC/VIUF Conference, 1998, Santa Clara, CA

[12] W. Putzke-Röming: *Foundations of Object-oriented Communication Between Structural Hardware Objects*, Technical Report, 1999

[13] M. Radetzki, W. Putzke-Röming, W. Nebel: *Objective VHDL: Tools and Applications*, Proceedings of First International Forum on Design Languages (FDL) 1998, Lausanne

[14] M. Radetzki: *Synthesis of Digital Circuits from Object-Oriented Specifications*, Dissertation at University of Oldenburg, 2000

[15] J. A. Rowson, A. Sangiovanni-Vincentelli: *Interface-Based Design*, Proceedings of DAC 1997, Anaheim, CA

[16] S. Swamy, A. Molin, B. Convot: *OO-VHDL Object-Oriented Extensions to VHDL*, IEEE Computer, October 1995

[17] VSI Alliance: *VSI System Level Design Model Taxonomy*, VSI Reference Document, Version 1.0, 25 October 1998

[18] P. Wegner, S. B. Zdonik: *Inheritance as an Incremental Modification Mechanism or What Like Is and Isn't Like*, Proceedings of European Conference on Object-Oriented Programming (ECOOP), Oslo, Norway 1988, pp. 53-77.

[19] P. Wegner: *Dimensions of Object-Based Language Design*. Proceedings of OOPSLA Conference, 1987

HIGH-LEVEL SYNTHESIS THROUGH TRANSFORMING VHDL MODELS

Anatoly Prihozhy
Belarusian State Polytechnic

Abstract   In this paper a method of transforming a behavioral VHDL-model to a functionally equivalent model with one basic block is proposed. High-level synthesis techniques including scheduling, allocation, and binding are modified for the model. These reduce the number of control steps, FSM states, state transitions, functional and storage units in an RTL-structure.

Keywords   High-level synthesis, VHDL, behavioral model transformation, scheduling, allocation, binding, design space exploration.

1.   INTRODUCTION

Usually, the designer starts by describing the required behavior and specifying the optimization goal and design constraints. Capturing this input, a high-level synthesis tool generates a structure at the register transfer level (RTL) [1-5]. If the designer is not satisfied with the synthesis result, she/he can modify the behavioral description, optimization criterion, or design constraints and try again. Nowadays the path-based high-level synthesis method is widely and successfully used in efficient synthesis tools [1-5,10,11]. It constitutes a theoretical basis for performing data-control flow graph (DCFG) analysis, scheduling, allocation, and binding for the whole behavior. At the same time, the method has the drawbacks as follows:

- The number of paths on the CFG can increase exponentially depending on the number of nodes (up to millions of paths are possible for real designs [10]).
- Reordering of statements in a path is not allowed.
- During scheduling, allocation and binding very complex combinatorial tasks are executed for each path and the obtained results are combined.

In this paper we prove that any behavioral description can be equivalently transformed to a one basic block model (OBBM) allowing more powerful synthesis results than the traditional approaches allow. Section 2 presents the idea of transformational synthesis. In Sections 3 and 4 transformation rules and techniques for inferring OBBM are described. Scheduling, allocation, and binding techniques for OBBM are presented in Sections 5 and 6. Section 7 gives some results.

135

P.J. Ashenden et al. (eds.), System-on-Chip Methodologies & Design Languages, 135–146.
© 2001 *Kluwer Academic Publishers. Printed in the Netherlands.*

2. TRANSFORMATION-BASED SYNTHESIS

Starting from a behavioral VHDL-description, we generate a number of equivalent descriptions [6-9] by applying transformations (Fig.1). The technique extends the set of design alternatives and supports an efficient design space exploration. In this paper we propose a method of transforming a behavioral VHDL-model to a functionally equivalent model constructed of one basic block. The existing scheduling, allocation and binding techniques can't be directly applied to the model. We modify the techniques by means of using several relations on the sets of variables and statements in order to improve parameters of the generated RTL-structure.

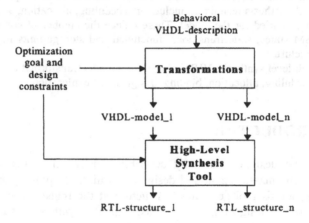

Figure 1. Synthesis through transformation

3. TRANSFORMATION OF ORIGINAL MODEL

We transform a behavioral sequential synchronous VHDL model with flexible or specified cycle behavior in order to obtain a VHDL-model that is more suitable for efficient high-level synthesis. Several types of useful transformations are used in the existing HLS tools [1-11]. We propose deeper transformations leading to significant reorganization of the source VHDL-model and its CFG: splitting, inserting, and attaching statements, eliminating loop-, exit- and next-statements, and others. We avoid multiple assignments to the same variable. Through the paper the following notations will be used: $V,V1,...,C,C1,...$ are Boolean variables and expressions, $Q,Q1,...$ are sequential statements, $S,S1,...$ are sequences of statements, and $L,L1,...$ are labels of loops.

Our transformation rules are significantly based on using loop-statements without an iteration scheme. The VHDL loop-statements *while* and *for* with ascending and descending ranges as follows:

while C **loop** S **end loop;**
for I **in** L **to** R **loop** S **end loop;**   (1)
for I **in** L **downto** R **loop** S **end loop**

may be replaced with

loop V:=**not** C; **exit when** V; S **end loop;**
I:=L; **loop** V:=I>R; **exit when** V; S I:=I+1; **end loop;**   (2)
I:=L; **loop** V:=I<R; **exit when** V; S I:=I-1; **end loop;**

   If an exit-statement is under an if-statement:

if C1 **then exit** L **when** C2; **end if;**   (3)

then the two statements may be merged as

V:=C1 **and** C2; **exit** L **when** V;   (4)

   If an exit-statement *exit L when C; S* is followed by a sequence S of statements that does not update the value of C, then the two statements may be reordered as:

if not C **then** S **end if; exit** L **when** C;   (5)

   Two exit-statements with the same loop-label:

exit L **when** C1; **exit** L **when** C2;   (6)

may be merged as:

V:=C1 **or** C2; **exit** L **when** V;   (7)

   All the next-statements are eliminated from the VHDL-behavioral description. The most general situation with a next-statement and two nested loops can be represented as:

L1: **loop** S1 L2: **loop** S2   (8)
next L1 **when** C; S3 **end loop** L2; S4 **end loop** L1;

Replacing the next-statement with exit- and if-statements yields:

V:=false; L1: **loop** S1 L2: **loop** S2 V:=C;   (9)
exit L2 **when** V; S3 **end loop** L2;
if not V **then** S4 **end if; end loop** L1;

   The same transformation can be performed on an arbitrary number of nested loops. It can imply introducing a label for an unlabeled loop.
   In VHDL an if-statement

if C1 **then** S1 **elsif** C2 **then** S2 ● ● ●   (10)
elsif Cn-1 **then** Sn-1 **else** Sn **end if;**

may be split to the following sequence of variable assignment and if-then-statements:

V1:=C1; V2:=**not** C1 **and** C2; ● ● ●   (11)
Vn-1:=**not** C1 **and not** C2 **and** ... **and** Cn-1;
Vn:=**not** C1 **and not** C2 **and** ... **and not** Cn-1;
if V1 **then** S1 **end if**; **if** V2 **then** S2 **end if**; ● ● ●
if Vn-1 **then** Sn-1 **end if**; **if** Vn **then** Sn **end if**;

In special case *if V1 then V2:=E; end if;* the if-then-statement may be replaced with a variable assignment statement:

V2:=(V1 **and** E) **or** (**not** V1 **and** V2);   (12)

In VHDL the case-statement has the following form:

case E **is when** H11|...|H1k$_1$ => S1 ● ● ●   (13)
when Hr1|...|Hrk$_r$ => Sr **when others** => Sr+1 **end case**;

where E is an expression of a discrete type and *Hij* is a choice defining a value or a range of values. The case-statement is equivalently transformed to the following if-statement:

V:=E; **if** R11 **or** ... **or** R1k$_1$ **then** S1 ● ● ●   (14)
elsif Rr **or** ... **or** Rrk$_r$ **then** Sr **else** Sr+1 **end if**;

where V is a variable of the type associated with the expression E and *Rij* is a relational operator (expression) associated with the choice *Hij*. The if-statement is split as it was described above.

The return-statements are used to exit from VHDL functions and procedures. We eliminate the statements from models and replace them with variable assignment, exit- and if-then-statements. If a signal or variable assignment statement contains an expression E in the right part constructed of more than one operator, it is split implying the addition of variables of appropriate types. Two approaches to processing the subprograms during high-level synthesis are possible: performing high-level synthesis separately for each subprogram and merging the process and subprograms before synthesis. In order to unify the merging algorithm, all the functions are replaced with procedures and assignment statements $U:=F(P);$ are replaced with procedure calls $F(P,U)$ where U is a variable, F is a function name, and P is a list of actual parameters.

4.   VHDL-MODEL WITH ONE BASIC BLOCK

The number of paths on CFG significantly depends on the number of basic blocks. In this section we prove that any VHDL-process may be represented as one basic block in the statement part:

P: **process** Declars   (15)
begin Basic_Block **end process** P;

where *Declars* are local declarations and *Basic_Block* is a sequence of wait, variable and signal assignment statements either covered or not covered with if-then-statements. An if-then-statement describes a condition of operation execution and variable/signal assignment. The process parenthesis *"begin"* and *"end"* describe an infinite loop. The proposed transformation technique consists of the key steps as follows: inserting statements located after a loop into the loop, inserting statements located before a loop into the loop, merging neighbor nested loops, eliminating loop- and exit-statements, and eliminating subprogram calls.

We assume that all the loop-statements are labeled and all the exit-statements refer a loop-label. If a sequence of statements is located after a loop as in the following fragment

L: **loop** S1 **exit** L **when** V; S2 **end loop** L; S3　　　　　　　　(16)

where S3 does not update the value of variable V, then inserting the statements into the loop yields:

L: **loop** S1 **if** V **then** S3 **end if**; **exit** L **when** V; S2　　　　　(17)
end loop L;

When statements located before a loop covered by an if-statement are inserted into the loop:

S1 **if** C **then** L: **loop** S2 **end loop** L; **end if**;　　　　　　　　(18)

then the inserting yields:

V:=true; L: **loop if** V **then** S1 **end if**;　　　　　　　　　　　(19)
V1:=V **and not** C; V:=false; **exit when** V1; S2 **end loop** L;

Usually a VHDL-process or subprogram contains a hierarchy of loops. In the hierarchy pairs of loops exist that either execute sequentially or one loop is in the body of other loop. For both cases an equivalent system of nested loops may be constructed. The transformations are based on the following two rules. First, a hierarchy of two loops:

L1: **loop** S1 **if** C1 **then** L2: **loop** S2 **exit** L2 **when** C2;　　　(20)
S3 **end loop** L2; **end if**; S4 **end loop** L1;

may be equivalently transformed to the following nested loops:

L1: **loop** V1:=true; L2: **loop**　　　　　　　　　　　　　　(21)
　　if V1 **then** S1 **end if**; V2:=V1 **and not** C1;
　　V1:=false; V3:=**not** V2; **if** V3 **then** S2 **end if**;
　　V4:=V3 **and** C2; **if** V4 **then** S4 **end if**;
　　V5:=V2 **or** C2; **exit** L2 **when** V5; S3
end loop L2; **end loop** L1;

Second, a sequence of two loops:

L1: **loop** S1 **exit** L1 **when** C1; S2 **end loop** L1;　　　　　　(22)

S3 L2: **loop** S4 **exit** L2 **when** C2; S5 **end loop** L2;

may be transformed to the nested loops as follows:

L1: **loop** V1:=true; L2: **loop**　　　　　　　　　　　　　　　　(23)
　　if V1 **then** S1 **end if**; V2:=V1 **and** C1;
　　if V2 **then** S3 **end if**; V3:=V1 **and not** C1;
　　V1:=false; V4:=**not** V3; **if** V4 **then** S4 **end if**;
　　V5:=V3 **or** C2; V6:=V5 **and** C1; **exit** L1 **when** V6;
　　if V5 **then** S2 **end if**; **exit** L2 **when** V5; S5
end loop L2; **end loop** L1;

　　Two nested neighbor loops:

L1: **loop** V1:=true; L2: **loop** S1 **exit** L2 **when** C1;　　　(24)
S2 **end loop** L2; **end loop** L1;

may be replaced with one loop:

L1: **loop** S1 **if** C1 **then** V1:=true; **end if**;　　　　　　(25)
V2:=**not** C1; **if** V2 **then** S2 **end if**; **end loop** L1;

without the exit statement referring *L2*. Given a process with *k* nested loops, the loops may be eliminated step by step. Finally, a process with one basic block is inferred. The process statement part is a loop suspended and resumed by events on the *Clock* signal.

　　We assume that a subprogram is called for more than once in a process body. Our goal is to reconstruct the process body in such a way as to attach only one copy of the subprogram body. The goal is achieved through using additional control Boolean variables and if-statements. We illustrate our approach with the following abstract VHDL-like fragment:

process ● ● ●　　　　　　　　　　　　　　　　　　　　　　(26)
　procedure F(Pi, Po) **is** Declars **begin** B **end** F;
begin S1 F(P1i,P1o); S2 F(P2i,P2o); S3 **end process**;

where *F* is a procedure name, *Pi* and *Po* are descriptions of input and output formal parameters, *Declars* is a list of declarative items, *B* is a procedure body represented as one basic block, and *P1i, P1o, P2i,* and *P2o* are input and output actual parameters of the first and second procedure calls. After inserting the declarations *Pi, Po, and Declars* into the process declarative part, inserting the procedure body *B* into the process statement part, and reorganizing the statement part, we obtain the fragment as follows:

process ● ● ● Pi, Po, and Declars modified　　　　　　　　(27)
　variable V1,...,V7: Boolean; **begin**
　　V3:=V1 **or** V2; **if** V3 **then** B **end if**;
　　V4:=**not** V3; **if** V4 **then** S1 **end if**;
　　if V4 **then** Pi':=P1i; **end if**; V5:=V1 **and not** V2;

if V5 **then** P1o:=Po'; **end if**; **if** V5 **then** S2 **end if**;
if V5 **then** Pi':=P2i; **end if**; V6:=**not** V1 **and** V2;
if V6 **then** P2o:=Po'; **end if**; **if** V6 **then** S3 **end if**;
V7:=V1 **xor** V2; V2:=V1; V1:=**not** V7;
end process;

where *Pi'* and *Po'* are local variables representing the input and output formal parameters of procedure. In the process, all variables that are used in *B* have to be unique. Increase in the number of procedure calls requires additional control variables, although the transformation method remains the same.

Now we demonstrate the proposed transformation technique on a simple VHDL behavioral description. An original GCD algorithm is presented in Fig.2. First, we split all the control structures and remove the iteration scheme from the loop. Then, we insert the statements that precede and succeed the loop into the loop. Additional variables *C0, C2,* and *C3* are used. Finally, the loop is eliminated, the exit-statement is replaced with an if-statement, and the variable *C0* takes initial value *true* (Fig.3).

5.   SCHEDULING FOR ONE BASIC BLOCK MODEL

Scheduling aims at introducing control steps and FSM states, distributing statements on the steps, and either minimizing the number of

```
entity GCD is
    port(Clock, Reset: in Bit; XP, YP: in Bit_Vector(15 downto 0);
      Ready: out Bit; Res: out Bit_Vector(15 downto 0));
    end GCD;
    architecture Behavior1 of GCD is  begin
    process  variable X,Y: Bit_Vector(15 downto 0); begin
      wait until Clock'Event and Clock='1'; Ready<='0'; X:=XP; Y:=YP;
      while (X/=Y) loop  wait until Clock'Event and Clock='1';
       if (X<Y) then Y:=Y-X; else X:=X-Y; end if;
      end loop; Ready<='1'; Res<=X;  end process;
     end Behavior1;
```

Figure 2. Original VHDL-model of GCD

```
architecture Behavior5 of GCD is  begin  process
    variable X,Y: Bit_Vector(15 downto 0);  variable C0: Boolean:=true;
    variable C2,C3: Boolean;  begin
    if C0 then wait until Clock'Event and Clock='1'; end if;
    if C0 then Ready<='0'; end if;  if C0 then X:=XP; end if; if C0 then Y:=YP;
    end if;  C0:=X=Y; if C0 then Ready<='1'; end if; if C0 then Res<=X; end if;
    if not C0 then wait until Clock'Event and Clock='1'; end if; C2:=X<Y;
    C3:=X>Y;  if C2 then Y:=Y-X; end if;
    if C3 then X:=X-Y; end if;  end process;
   end Behavior5;
```

Figure 3. Transformed VHDL-model of GCD

steps or minimizing the design cost. The scheduling results in generating a high-level state machine (HLSM). The known scheduling techniques which process a traditional basic block are [1-5]: as soon as possible (ASAP), as late as possible (ALAP), list scheduling, integer linear programming formulation (ILPF), freedom-based scheduling, force-directed scheduling, dynamic loop scheduling, and scheduling for pipelines. The path-based scheduling technique processes the whole CFG consisting of several basic blocks.

In a traditional basic block all statements execute unconditionally. In the OBBM a statement may execute conditionally or unconditionally. Our scheduling methodology includes: analyzing data dependencies, generating orthogonal, implication, and equivalence relations on the set of control signals and variables, generating a set of pairs of orthogonal statements, analyzing operator compatibility and operator proximity, generating a statement precedence relation, performing a modified scheduling technique (ILPF, list scheduling, ASAP, and ALAP). The generated schedule is optimized due to orthogonal operations are mutually exclusive and may execute on the same functional unit concurrently without additional resources. As a result, the statements are reordered, grouped, and an optimized HLSM is generated.

Signals and variables of Boolean type control the execution of if-then-statements. Relations on the set of control variables define in which manner the operators and assignments that are under an if-then-statement execute. Two control variables $V1$ and $V2$ are:

- Orthogonal (#) if the variables never take value *true* simultaneously
- Variable $V1$ implies variable $V2$ ($\rightarrow$) if $V2$ takes never value *false* when $V1$ equals *true*
- Equivalent ($\leftrightarrow$) if the variables can't take different values simultaneously
- Independent (−) if neither of the previous cases takes place.

There are two main sources for inferring the relations. First, the VHDL relational operators are analyzed. Given two assignment statements $V1:=X1$ $R1$ $Y1;$ and $V2:=X2$ $R2$ $Y2;$ with the relational operators $R1,R2 \in \{=, /=, <, <=, >, >=\}$ are, the relations #, $\rightarrow$, $\leftarrow$, and $\leftrightarrow$ between $V1$ and $V2$ are inferred through using Table 1. Second, the VHDL logical operators are analyzed (Table 2) to infer additional relations between other pairs of control variables. In the table, A, B, C, and D are Boolean variables. There are similar rules for pairs of variable assignments and for pairs of relations.

All the sequential statements in OBBM may be considered as an if-then-statement. If a statement Q is not conditional one, it may be replaced with *if true then Q end if;*. Two if-then-statements *if V1 then Q1 end if;* and *if V2 then Q2 end if;* are defined to be orthogonal if the variables $V1$ and $V2$ are orthogonal. One orthogonal statement cannot precede another orthogonal statement. The orthogonal statement bodies are mutually exclusive and may execute on the same functional unit concurrently in the same control step.

Table 1   Inferring relations between *V1* and *V2*

N	Operator *R1*	Operator *R2*	Relation between *V1* and *V2*
1	=	/=, <, >	#
2	/=	<, >	←
3	<	=, >, >=	#
4	<=	=, <	←
5	>	=, <, <=	#
6	>=	=, >	←
and others			

Table 2   Inferring relations using logical operators

Statement	Existing relations	Inferred relations
A:=B **and** C;	D#B	D#A
A:=B **or** C;	D#B and D#C	D#A
A:=B **xor** C;	B#C	B→A and C→A
A:=B **nand** C;	D→B and D→C	A#D
A:=B **nor** C;	–	A#B and A#C
A:=**not** B;	–	A#B
and others		

There are two cases for operators to be compatible within one control step: the operators belong to orthogonal statements and may execute on the same type of functional unit and the operators are relational and have identical operands. The compatible operators may execute within one control step without additional resources. The operator proximity is used to select compatible operators to be merged. We estimate the proximity of two operators as the number of common inputs and outputs. Maximizing the proximity of operators leads to minimizing the cost of interconnection units in resulting RTL-structure.

The statement precedence relation $PRE = VL \cup US \cup WT$ is a union of three sub-relations:

- A statement i precedes a statement j $((i,j) \in VL)$ if i and j are not orthogonal and i has an output variable that is an input variable for j
- A statement i precedes a statement j $((i,j) \in US)$ if i and j are not orthogonal and i uses a value of an input variable to be assigned a new value by j
- A statement i precedes a statement j $((i,j) \in WT)$ if a wait-statement w exists such that the pairs (i,w) and (w,j) of statements are not orthogonal and i precedes w and j succeeds w in the VHDL-text.

The relation PRE can be represented as a statement precedence graph. Statements i and j are sequential if a path exists between i and j in the graph, otherwise, the statements are concurrent.

All the traditional techniques [1-11] scheduling a basic block assume that the sequential statements may not be an if-then-statement and, therefore, may not be orthogonal. Our method allows the orthogonal statements to execute in the same control step in parallel without additional resources. This implies increase in the average number of statements in a step and decrease in the total number of steps. As a result, the number of values computed in one control step and used in another step is reduced as well as the number of registers. The extended ASAP and ALAP being feasible-constrained techniques use the statement precedence graph at input. The modified list scheduling being a resource-constrained technique uses a status of each statement: (1) all the predecessors have been already scheduled, (2) there is a predecessor has not been scheduled, (3) the statement may be scheduled on an existing functional unit using the orthogonal relation, (4) addition of a functional unit is required, and others. The modified time- and resource-constrained scheduling problems ILPF can be formulated based on work [4].

The HLSM initially includes a sequence of states and a set of control variables (Fig.4). Each state has exactly one transition. To speed up the HLSM operation, additional direct transitions are added.

6.   ALLOCATION AND BINDING FOR OBBM

The RTL-structure being an output of high-level synthesis consists of two parts: a data path (DP) and a finite state machine (FSM). An allocation aims at minimizing the DP cost and defining the set of functional, storage, and interconnection units in the RTL-structure. A binding aims at mapping the elements of behavioral description to the structure components. A lot of allocation and binding techniques have been developed [1-7]: path-based, rule-based, branch and bound, clique partitioning, integer linear programming, simulated-annealing, graph coloring, and other algorithms. The OBBM being a purely data flow representation supports an efficient allocation and binding: analysis of variable lifetimes in one basic block, generating the variables compatibility relation, generating the operators compatibility and proximity relations, deriving the DP from the DFG and folding it.

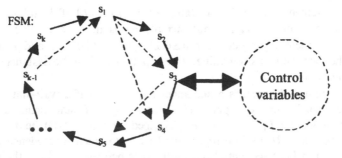

Figure 4. Control part of RTL-structure

Assuming that all the statements are distributed on the sequence of HLSM states, a variable v lifetime is represented as an interval $l_v=[s^b_v, s^e_v]$ where s^b_v, and s^e_v are the earliest and latest states in which the variable is alive. In order to compute the interval we use the function $Inc: V \times S \rightarrow \{\varnothing, \{in\}, \{out\}, \{in, out\}\}$ mapping the pairs variable/state to subsets of the set $\{in, out\}$. It is easy to derive from the function the first state s_v^{first} in which v is used as output and the last state s_v^{last} in which v is used as input. The values s^b_v, and s^e_v are computed from the values s_v^{first} and s_v^{last} taking into account the fact that the basic block is a body of an infinite loop.

A variable $v \in V_W$ which is alive within one HLSM state is implemented as a wire. A variable $v \in V_M$ which is alive in several HLSM states is mapped to a register, RAM, or ROM. Two variables v_1 and v_2 of V_M may be mapped to the same storage unit if they are compatible. The variables are compatible if either their lifetime intervals are not intersected or each statement using v_1 is orthogonal to each statement using v_2. We describe the variable compatibility as a binary relation C_V. The operator compatibility is represented by a binary relation C_O computed using the formula:

$$C_O=(U_O \frown (S_O \backslash O_O)) \cup R_O,$$ (28)

where '$\backslash$' is a set subtraction operation; $\sim A$ is complementation of set A; U_O is the set of pairs of operators executed on the same type of functional unit; S_O is the set of pairs of operators executed in the same HLSM state; O_O is the orthogonal relation; R_O is the set of pairs of relational operators executed in the same HLSM state and using the same operands. The operator proximity relation P_O is estimated as it was described in Section 5.

Folding the DFG and DP aims at minimizing the design cost. Two types of optimization algorithms have been developed: (1) global optimization: the variables, operators, and data dependencies are merged in parallel, (2) local optimization: the variables, operators, and data dependencies are merged separately. The algorithms of the first type find a cheaper design, while the algorithms of the second type are faster.

7. RESULTS

A software which implements the high-level synthesis techniques based on transformation of VHDL-models has been developed. Some advantages and drawbacks of the proposed method are listed in Table 3. Table 4 presents experimental results for three VHDL-models of the Bubble benchmark [5]. In model 1 all the expressions and conditional statements are split and the iteration scheme of loops is removed. In model 2 the statements located between and after loops are inserted into the loops. In model 3 the loop- and exit-statements are eliminated. The number of control steps is decreased twice due to the transformations. A non-significant increase in the register cost is due to scheduling for minimal register count has not been used.

Table 3	Advantages and drawbacks of OBBM
Advantages	**Drawbacks**
Reduction in the number of: • Basic blocks in CFG • Control steps and FSM states • State transitions • Storage and functional units	The transformed behavioral VHDL-model differs from the initial VHDL-specification
More opportunities for pipelining, chaining, multi-cycling, and asynchronous high-level synthesis	Extending the set of control variables

Table 4			Bubble benchmark
Parameter	Behavioral VHDL-model		
	Bub_1	Bub_2	Bub_3
Loops	9	9	0
Control steps	20	15	10
Registers	7	9	10
Register width (bits)	104	106	107
Multiplexers	4	4	6
Multiplexer width (bits)	68	68	70
RAM	1	1	1

8. REFERENCES

[1] R.A.Bergamaschi, *"High-Level Synthesis in a Production Environment"*, Fundamentals and Standards in Hardware Description Languages, J.P. Mermet, ed., Kluwer Academic Publishers, Norwell, Mass., 1993, pp.195-230.

[2] R.Camposano and W.Rosensteil, *"Synthesizing Circuits from Behavioral Descriptions"*, IEEE Trans. CAD, Vol.CAD-8, Feb. 1989, pp.171-180.

[3] D.D.Gajski et al., *"High-Level Synthesis: Introduction to Chip and System Design"*, Kluwer Academic Publishers, Norwell, Mass., 1992.

[4] T.Hwang, J.Lee, Y.Hsu, "A Formal Approach to the Scheduling Problem in High- Level Synthesis", *IEEE Trans.on CAD*, Vol.10, No.4, 1991.

[5] A.A.Jerraya, I.Park, and K.O'Brien, *"Amical: An Interactive High-Level Synthesis Environment"*, Proc. European Design Automation Conf.93, IEEE Computer Society Press, Los Alamitos, Calif., 1993.

[6] A.Prihozhy, *"Net Scheduling in High-Level Synthesis"*, IEEE Design & Test of Computers, Spring, 1996, pp.26-35.

[7] A.Prihozhy "Asynchronous Scheduling and Allocat-ion", Proc. DATE 98, *IEEE CS Press*, CA, 1998.

[8] A.Prihozhy and F.Buijs *"Transformations of Behavioral VHDL-Descriptions"*, National Academy of Sciences, Belarus, 1994.

[9] A.Prihozhy, *"Methods for Logical Algorithm Equivalent Transformation in VLSI CAD"*, Trans. Physics & Mathematics, National Academy Sciences, Belarus, 1992, N 2, pp.86-92.

[10] W.Rosensteil, *"Experiences with High-Level Synthesis from VHDL-Specifications"*, Proc. Workshop on Design Methodologies for Microelectronics and Signal Processing, Gliwice-Cracow, 1993, pp.405-412.

[11] E.Villar and P.Sanches, *"Synthesis Applications of VHDL"*, Fundamentals and Standards in Hardware Description Languages, J.P. Mermet, ed., Kluwer Academic Publishers, 1993.

SPECIFICATION FORMALISMS

Multi-Facetted Modeling
The Key to Systems Engineering

PERRY ALEXANDER
The University of Kansas

DAVID BARTON
AverStar, Inc

Key words: Systems level design, Rosetta, facets, domains, modelling

Abstract: Systems engineering involves the integration of information from many
disciplines into a design activity. To achieve this end, it is necessary to
provide system design representations from multiple perspectives using
multiple modeling paradigms. The systems-on-a-chip problem is no different,
implying that any Systems Level Design Language (SLDL) used in this area
must support multi-faceted, heterogeneous design. This paper overviews
successive attempts to support systems level design using VHDL, VSPEC
and Rosetta.

1.   INTRODUCTION

Systems level design is characterized by the need to deal with
heterogeneity during the design process. Heterogeneity arises from two
sources: (i) heterogeneous component use; (ii) heterogeneous component
models. In large, heterogeneous systems, different components are best
modeled using different semantic models. Digital electronic, analog
electronic, optical, and MEMS components all have different underlying
mathematical domain models. Multiple aspects, or facets, of the same
component are best modeled using different underlying semantic models.
Electromagnetic, analog, digital and constraint facets have different
underlying mathematical domain models.

149

P.J. Ashenden et al. (eds.), System-on-Chip Methodologies & Design Languages, 149–159.
© *2001 Kluwer Academic Publishers. Printed in the Netherlands.*

Any systems level modeling language must include: (i) multiple modeling paradigms for different component facets; (ii) multiple modeling paradigms for different component domains; and (iii) a means for moving information between system representations. Multiple modeling paradigms supports integrated modeling. Moving information between system representations supports using multiple semantic models without forcing a single model.

This paper presents a brief history of efforts to provide systems level modeling capabilities. As VHDL served as the starting point for our modeling capabilities, the paper begins with a brief description of its systems level modeling capabilities. Then, a brief overview of VSPEC is presented. VSPEC is an interface specification language for VHDL that represents an initial attempt to model multiple component facets at the requirements level. Specifically, providing declarative functional and performance constraint modeling capabilities. Finally, the emerging systems level description language Rosetta is presented with an overview of its modelling characteristics and design goals.

2.   STARTING WITH VHDL

VHDL provides users with both behavioral and structural modeling capabilities for digital systems. It provides an operational language for describing the behavior of a component. This language subset, referred to as *behavioral* VHDL, allows users to describe data transforms and control structures for components using a traditional operational style. VHDL also provides users with a declarative language for describing the structure of a system. This language subset, referred to as *structural* VHDL, allows users to describe interconnections between components using a net list-based module interconnect language.

By defining behavioral and structural architectures for the same component allows VHDL users model *behavioural* and *structural* facets of a component. Users provide a high level, black-box behavioral description and use that description as a basis for refinement into a structural system definition. Such activities are common in top-down design processes making these facets and their interaction quite useful to systems designers.

Behavioral and structural VHDL share a common simulation-based semantics that supports mixing facets and information. More specifically, the results of simulating structural and behavioral representations of a component can be directly compared. When simulated, all VHDL descriptions are elaborated into a common executable subset. Thus, designers are able to evaluate design iterations by simulating results.

Further, models using different representational styles may be intermixed in a design activity.

Although VHDL has excellent operational specification capabilities, their application during the systems level design process is limited. A limitation noted in our research activities is at the requirements specification level. Specifically: (i) VHDL's operational semantics are not suitable for abstract functional requirements; and (ii) VHDL provides no means for describing performance requirements. These two information classes represent facets important in the systems design process. VSPEC is an initial attempt to address these information types in the context of VHDL.

3.   VSPEC - A FIRST STEP

VSPEC uses an axiomatic specification technique for representing a component's functional and performance requirements. A pre- and post-condition are defined to indicate: (i) what must be true in the current state (I); and (ii) what must be true in the next state (O). The pre- and post-condition follow the traditional axiomatic semantics presented by Hoare [1] and are implemented using a Larch Interface Language approach [2]. The axiomatic specification is augmented with an activation condition indicating what circumstances cause the component to activate. The activation condition is needed because of the concurrent nature of VHDL components in contrast to the serial nature of software components.

The traditional axiomatic approach is further modified to describe performance requirements. Each performance requirements is modeled using a declarative semantics to express relations over constraint variables. Although constraints are expressed declaratively, their semantics varies based on the information represented. This is particularly true when combining component constraints within an architecture.

3.1   An Example

An example of a simple VSPEC search component is presented in Figure~\ref{fig:example}. This component accepts an array of elements and a key and returns the array element associated with that key. Changing the value of either the key input or the array to be searched causes the component to activate.

```
package search_types is
    type E is mutable;
    type K is mutable;
    type E_array is array (integer range <>) of E;
```

```
end search_types;

use work.search_types;
entity search is port
   (input: in E_array; k: in K; output: out E);
   includes KeyToElement(E, K), Area, Power, Frequency;
   modifies output;
   sensitive to
      k'event or input'event;
   requires true;
   ensures forall e: E
      ((output'post = e) iff (key(e)=k and e in input));
   constrained by
      area <= (3 um * 5 um)
      and power <= 10 mW
      and clock_frequency <= 50 MHz;
end search;
```

Figure 1. An example VSPEC specification defining requirements for a simple search component.

3.2   Functional Requirements

The basic specification model used for VSPEC functional requirements is a state machine. Using the axiomatic style, a state machine is specified defining interface characteristics of the component. Pre-conditions and post-conditions define the output and next state functions while entity ports and VSPEC state variables define component state. Performance constraints must be satisfied in every state, however their semantics are not necessarily state based.

Functional requirements are modeled using the activation condition, pre-condition and post-condition. These are specified in the sensitive to, requires, and ensures clauses, respectively. The requires clause defines a pre-condition that must be true in the current state for the component to execute correctly. The ensures clause defines a post-condition the component must make true in the next state given the pre-condition is true in the current state. Given that x is the collection of entity ports and VSPEC state variables providing input or state and z is the collection of entity ports and VSPEC state variables providing output or next state, the relationship defined by the requires and ensures clauses can be defined as:

$$\forall\, x \cdot requires(x) \Rightarrow \exists\, z \cdot ensures\ (x,z)$$

The axiomatic specifications define the data transformation performed by the component. These specifications define requirements on how the component transforms data by defining relations between inputs, current state and outputs. Specifically, any implementation of these requirements, F(x), must provide a witness for z that satisfies the requirements. Skolemizing Equation~\ref{eq:axiomatic} results in the following correctness condition for the data transformation:

$$\forall x \; \text{requires}(x) \Rightarrow \text{ensures}(x, F(z))$$

The importance of this relationship is the connection it provides between the requirements defined by VSPEC and the execution of a VHDL implementation. Given only these requirements, any VHDL implementation of F(x) is a correct implementation. Thus, the requirements facet is connected semantically to the behavioral or structural facet. Further, the requirements facet could be associated with a test facet or other facet defined for a component.

The activation condition defined in the sensitive to clause defines when a component becomes active. Like the pre-condition, the activation condition must be true for the component to function. If the pre-condition is false, the component's behavior is undefined. In contrast, if the activation condition is false the component maintains its current state. The activation condition models events that cause the component to perform its task.

When components are interconnected, activation conditions model interaction between components. Activation conditions are defined over the same symbol set as pre-conditions. They monitor inputs and state to determine when the component should perform its data transform. When inputs are connected to outputs from other components or inputs from outside the system, control is communicated between components. Activation conditions are modeled using a process algebra. Process algebras are designed specifically to model reactive systems and suit the semantic needs of activation conditions nicely. Specifically, VSPEC activation conditions are modeled using CSP [3].

Each VSPEC entity is modeled as a CSP process. The alphabet of each process' is the set of system states where its associated activation condition is true. By definition, the CSP process ignores any symbol not in its alphabet. Thus, any state where the component is not active is ignored by the component.

3.3   Architectures

An architecture is a collection of interacting components. VHDL provides structural descriptions for defining component and process interconnection. VSPEC uses the same structural descriptions to connect entities annotated with VSPEC definitions. A VSPEC structural description is exactly analogous to structural architectures used in traditional VHDL. Figure~\ref{fig:architecture} shows a VSPEC component architecture for a \texttt{search} architecture. This architecture specifies a sort component that prepares input for a binary search component. Figure~\ref{fig:components} defines the VSPEC requirements for the components used in the architecture.

```
use work.search_types;
architecture structure of search is
   component sort
     port (list_in: in E_array;
           list_out: out E_array);
   end component;
   component bin_search
     port (list_in: in E_array;
           k: in K;
           e: out E);
   end component;
   signal x: E_array;
 begin
   C1: sort port map (input,x);
   C2: bin_search port map (x,k,output);
 end structure;
```

Figure 2. A candidate architecture for the search example

VSPEC's process algebra semantics supports defining bisimulation relationships [4] between single component requirements and VSPEC component architectures. A VSPEC architecture is a decomposition of a system into a collection of interconnected components. The requirements of each component are known but an implementation has not yet been defined. A VSPEC architecture represents a decomposition step in a top down design process. Bisimulation relationships define when a VSPEC architecture exhibits behavioral equivalence with its associated requirements; *e.g.* when they look the same at their interfaces. Using the axiomatic semantics of VSPEC with its process algebra control semantics a structural facet (the VSPEC architecture) can be related with a requirements facet (the component specification).

Interaction between a VSPEC architecture and constraints uses a different semantics for the architecture. Interpreted with respect to heat dissipation, the architecture semantics implies an additive relationship between component constraints and the system level constraints. Similarly, pin-to-pin timing constraints are composed in an architecture using longest paths through components. What should be noted here is that the semantics of architecture (i.e. component composition) vary from semantic domain to semantic domain.

3.4 Performance Constraints

Performance requirements are modeled using relations defined in the constrained by clause. These relations define constraints over a collection of variables defining constraint types. The component is required to meet those constraints at all times in every state. Thus, constraints behave much like invariants with respect to the axiomatic functional requirements.

The semantics of constraints can be defined in terms of a component's state. Simply, the constraint predicate must be true in all states. In VSPEC, physical types behave like VHDL physical types. Thus, these relations define constraints on area, power consumption and clock speed.

Constraints present special problems when interacting with other facets. Requirements, behavior and structural facets interact in relatively intuitive ways. Providing proper semantics defines clean relationships between facets. Constraints do not share this characteristic. Constraint variables (*e.g.* heat and area) have no analog in any other facet. Further, it is difficult if not impossible to model the relationship between a functional requirement and a constraint. Constraints have neither a simulation or true axiomatic semantic making relationships difficult to define.

4. ROSETTA – THE NEXT STEP

The primary lesson from VSPEC is that adding and integrating heterogeneous models to system representations is an effective means for managing systems level problems. However, VSPEC is inherently limited in its extensibility and its ability to represent new specification domains. The Systems Level Design Language community is currently examining solutions and problems presented by a heterogeneous modeling language. Such a language provides modeling support for different design domains employing semantics and syntax appropriate for those domains. Unlike traditional single semantics languages, the Systems Level Design Language provides a collection of interacting domain theories for describing systems facets. Each domain theory provides a semantic and representational framework for one or more design domains. Domain theories include data, computation and communication models. Each facet is written using a domain theory and describes a particular aspect of a system. Using the design abstractions provided by its domain theory, a facet describes requirements, behavior, constraints, or function of a system. Domain models do not share a common semantics, but instead share information where necessary. Thus, each design facet is defined using design abstractions from its associated domain rather than forcing a common design model.

4.1 Facet Definition

A *facet* is the basic unit of specification in Rosetta. A facet is a parameterised construct used to encapsulate definitions. Each facet represents a view, or model of a component or system. Because facets represent models from different domains, each facet may use its own semantic basis when defining models. Specifically, each facet specifies what *domain* is associated with the model represented.

Facets are defined using two primary mechanisms: (i) direct definition; and (ii) composition from other facets. Direct definition is achieved using a traditional syntax similar to a procedure in a traditional programming language or a theory in an algebraic specification language. Facet composition represents representing a system with two models simultaneously. As an example, a specification for the sort-req component follows:

```
facet sort-req(x:: in sequence(T); z:: out sequence(T))
begin state-based
   l2: permutation(z',x);
   l1: ordered(x');
end sort-req;
```

This declaration defines a facet sort-req with parameters i and o as a sequence of elements of sort T. The begin-end pair delimits the domain specific terms within the facet. The begin statement opens the set of terms and identifies the semantic domain of those terms. In this case, the semantic domain is state-based. Specifying a semantic domain indicates what domain theory the facet is being defined in. Effectively, it includes other facets and systems that define its semantic basis. If no domain is specified, the terms are assumed to be facet specifications and facet operators. Terms are labeled, well formed formulas (wffs) with respect to the facet's semantic domain.

4.2   Facet Composition

Facet composition is the process of combining multiple facets into a single definition. In Rosetta, facet composition is achieved by defining facets and using logical connectives such as AND, OR and NOT to combine definitions. AND combines the properties of two facets and asserts that both sets of properties must hold simultaneously. Mathematically, AND defines a *product* of the two component models. OR combines the properties of two facets and asserts that at least one set of properties must hold. Mathematically, OR defines the *sum* if the two component models. NOT defines the negation of a component model and serves to define the implication operator and model equivalence.

An example serves to illustrate facet composition. We extend the sort example introduced previously. The definitions of facets sort-req and bubble-sort as defined above can be composed using conjunction to define a facet sort as:

```
sort = sort-req and bubble-sort;
```

The composition of sort-req and bubble-sort provides requirements and a postcondition for the sort routine. By defining the semantics of the operational facet such that termination of bubble-sort corresponds to the next state, the conditions placed on o' provide a postcondition requirement for the routine. Furthermore, the postcondition could provide an assertion for the sorting routine.

One could also have a temporal requirements specification sort-timed} The o@t notation in sort-timed refers to the value of o at time t.} and a constraints specification sort-const composed with the operational bubble-sort facet:

```
facet sort-timed(i :: in T[]; o :: out T[])
begin continuous;
  permutation(o@(t+5mS),i);
  ordered(o@(t+5mS));
end sort-timed;

facet sort-const
  power :: real;
begin constraints;
  p1: power <= 5mW;
end sort-const;

sort = sort-timed and sort-const
       and bubble-sort;
```

4.3   Facet Inclusion and Structure

Facet inclusion allows compositional definition in a manner similar to packages or modules in programming languages and theory inclusion in a formal specification language like PVS [5]. Whenever a facet label is referenced in a term, that facet may be included in the facet being defined. System structure is defined using facet inclusion and labeling. Facets representing components are included and interconnected by instantiating parameters with common objects. Labeling provides name spaced control and supports defining multiple instances of the same component. Consider the following specification of a two-bit-adder using two one-bit-adders:

```
facet one-bit-adder(x,y,cin :: in bit;
                    z,cout :: out bit)
  delay :: real;
begin requirements
  l1: z'= x xor y xor cin;
  l2: cout' = x and y;
end adder;

facet two-bit-adder(x0,x1 :: in bit;
                    y0,y2 :: in bit;
                    z0,z1,c :: out bit)
  delay :: real;
begin structural
  cx: bit
  b0: one-bit-adder(x0,y0,0,z0,cx);
  b1: one-bit-adder(x1,y1,cx,z1,c);
  delay = b0.delay+b1.delay;
end two-bit-adder;
```

Facet interconnection is achieved by sharing symbols between component instances. When a facet is included in the structural facet, formal parameters are instantiated with objects. When objects are shared in the parameter list of components in a structural facet, those components share the object. Thus, information associated with the object are shared between components. The two-bit-adder specification includes two copies of one-bit-adder. Parameters of the two adders are instantiated with parameters from two-bit-adder to associated signals with those at the interface. The internal variable cx is used to share the carry out value from the least significant bit adder with the carry in value from the most significant bit adder.

When the two one-bit-adder instances are included in the two-bit-adder definition, they are labeled with b0 and b1. The result is that the first `one-bit-adder` is renamed `b0` and the second `b1`. The implication of the renaming is that the physical variable `delay` associated with the adder definition is duplicated. Without renaming using labels, both `one-bit-adder` instances would refer to the same physical variable, `one-bit-adder.delay`. This is not appropriate as the adders should be distinct. The same result can be achieved using a parameter for delay. In large specifications including parameters for physical variables representing constraint specifications becomes cumbersome. Further, delay is not a parameter but a characteristic of the component.

5.   SUMMARY

Systems level design is a process of integrating information from multiple sources into the design decision making process. Any language attempting to address systems level design problems must address the representation and integration of heterogeneous component models. This paper presents an overview of three progressive attempts at addressing systems level design. VHDL provides a primitive systems level modelling capability by providing both structural and behavioural modelling capabilities. VSPEC extended this model by providing a requirements and performance constraint specification capability. However, both languages are limited by their extensibility and their adherence to a single semantic model. The emerging Rosetta systems level design language is presented as an alternative for representing and integrating heterogeneous systems models in component and system design models.

6.   REFERENCES

1. Hoare, C. A. R., "An Axiomatic Basis for Computer Programming," *Communications of the ACM*, **12**, 1969, pp. 576-580 ,583
2. Guttag, J. and Horning, J., Larch: Languages and Tools for Formal Specification," Springer-Verlag, New York, NY , 1993
3. Hoare, C. A. R., *Communicating Sequential Processes*, Prentice-Hall, Englewood Cliffs, NJ, 1985
4. Milner, R., *Communication and Concurrency*, Prentice Hall, Englewood Cliffs, NJ, 1989
5. Owre, S. and Rushby, J. and Shankar, N., "PVS: A Prototype Verification System," *Proc. of 11th International Conference on Automated Deduction, Lecture Notes in Artificial Intelligence* 607, 1992, pp. 748-752.

A Dual Spring System Case-Study Model in Rosetta

Peter J. Ashenden
Adelaide University

Perry Alexander
The University of Kansas

David L. Barton
Averstar

Key words: Rosetta, system level design, mechanical spring system, requirements, constraints.

Abstract: This paper describes a design case-study undertaken as part of the language design validation for the Rosetta System Level Description Language. The system under consideration is a dual-spring mechanical system. A physical model is constructed, describing the static characteristics of springs and the requirements and constraints applying to the system. It is demonstrated that the declarative modelling style used in Rosetta is a powerful modelling tool. Furthermore, the formal semantic basis of the language allows analysis of models and interfaces with other tools.

1. INTRODUCTION

Rosetta [1,2,3] is a system-level design language, being developed by two of the authors, Alexander and Barton, as part of the SLDL Initiative [4] currently sponsored by Accellera (formerly VHDL International) and ECSI. The SLDL Initiative was originally sponsored by the EDA Industry Council, and moved under the auspices of VI and ECSI in 1999. Rosetta addresses a need for a language in which designers can specify the requirements and constraints on a system that spans multiple design domains. Requirements describe functional behavior that a system must exhibit, and constraints describe operational bounds within which the system must remain. The different design domains include digital and analogue electronic subsystems, and the mechanical, optical, fluidic and thermal subsystems with which they interact.

P.J. Ashenden et al. (eds.), System-on-Chip Methodologies & Design Languages, 161–170.
© 2001 *Kluwer Academic Publishers. Printed in the Netherlands.*

Various computational models are most appropriate to describe aspects of systems in different domains. Example computational models include finite and infinite state-based, discrete event, discrete time, and continuous time. The Rosetta language is extensible to allow designs to be expressed using each of these computational models, and to allow expression of the interactions between descriptions in each computational model. The language achieves this extensibility by being based on a formal semantic underpinning and by including facilities for reflection. In combination, these features allow definition of syntax and semantics of new Rosetta domains for expressing requirements and constraints using different computational models.

This paper describes a model of a dual-spring mechanical system, developed as a design case study in Rosetta. The design was originally specified informally using a combination of English language and mathematical equations. The information in this specification formed the basis of a Rosetta specification, in which the requirements and constraints are expressed formally in a manner that is amenable to analysis and computation. The model is expressed in the *logic* domain in terms of continuous mathematical equations that describe the mechanical system.

2.   OVERVIEW OF THE ROSETTA LANGUAGE

The key language feature in Rosetta for describing a system is a *facet*. A facet captures one aspect of a system, describing properties of interest in one chosen domain. A given system may be described by a number of different facets, each characterising the system in a different domain. The various facets may be combined to provide a multi-faceted description, which might be instantiated as part of a larger system. In this section, we present a very brief overview of facets to enable the reader to understand the spring model that follows.

A facet, in its textual rendition, is a named unit comprising a parameter list, a collection of declarations, a domain identifier, and a collection of labelled terms. An illustrative example of a requirements facet is:

```
facet trigger(x::in real; y::out bit) is
    s:: bit;
begin continuous-time
    t1:   if s = 1 then
              if x >= 0.4 then s' = 1 else s' = 0 endif;
          else
              if x =< 0.7 then s' = 0 else s' = 1 endif;
          endif;
```

```
    t2: y@t+10ns = s;
end trigger;
```

The facet is named trigger, and represents the behavior of a Schmidt trigger device. The two parameters, x and y, each are of type bit. The parameter modes, in and out, are simply assertions whose meaning is defined in the domain of the facet. The domain is identified after the keyword begin, continuous-time in this example, representing the predefined continuous-time domain. The declaration before the keyword begin names the variable s and specifies its type as bit. The two terms t1 and t2 are assertions about the values of the parameters and the variables that must be true at all times. The variable t is implicitly defined in the continuous time domain to refer to the "current" time, and each term in the facet is implicitly universally quantified over all times. It is important to note that the operator "=" is an equality operator, not assignment. Thus the terms should be read as statements about what must be true at all times, rather than as operational definitions of events.

The notation s' denotes the value of s at the next instant after the current time. As shown in the example model, this allows description of discontinuities in the value of a variable over time. The notation y@t+10ns denotes the value of the variable y at a time 10ns later than the current time t. This allows expression of temporal relationships between values of variables.

The example model shown above uses simple bit-valued and real-valued types for variables. Rosetta has a very rich type system, including scalar discrete and continuous types, and type constructors for arrays, records, tuples and sequences. Furthermore, it permits definition of higher-order types, namely, types whose values are functions. The type system has well-founded mathematical semantics. This, in conjunction with the formal semantic definition of facets and their constituents, makes formal analysis and verification of Rosetta models feasible.

Facets, such as the one shown above, can be instantiated as terms in other facets, with variables or expressions being associated with the formal parameters. This provides a form of structural composition of models. As an illustration, the Schmidt trigger facet might be instantiated in a sensor model as follows:

```
facet sensor (...) is
   data_in:: real;
   quantized_data:: bit;
   ...
begin continuous
   s1: trigger(data_in,quantized_data);
```

```
    . . .
end sensor;
```

Furthermore, different facets of a given system can be composed using facet operators to form a description covering the different aspects of the system. One of the more common compositions is facet conjunction using the facet operator "and", which requires that all terms of both operand facets be true. Other composition operators include "or" for describing a system that may have several variants, "<=>" for asserting equivalence of two facets, and "=>" (implication) for asserting that one facet is a refinement of a more abstract facet. These and other composition operators are described in more detail in The Rosetta Usage Guide [1].

The domains referred to in the above descriptions are predefined sets of definitions that enlarge the syntactic and semantic vocabulary of the base language for particular kinds of modelling or computation. A domain definition consists of a collection of declarations and terms that are included in facets based on the domain. For example, the continuous facet used in the example model above defines the time variable t, and also defines the syntax and semantics of the "'" and "@" symbols. Rosetta predefines a number of domains that will be used in a wide variety of models. The fundamental domain is called logic, and includes definitions of the basic mathematical types, operators and expressions. Other predefined domains inherit from the logic domain.

Where a model is composed of facets from different domains, Rosetta *interactions* define the way in which terms in one domain imply properties in other domains. For example, an interaction that deals with conjunction of the logic domain and the continuous time domain might include a rule specifying that each term in the logic-domain facet is true for all time in the continuous-time domain.

While the Rosetta language is extensible through the definition of new domains and interactions, it is anticipated that most designers will only need to use the predefined domains and interactions. Development of new domains and interactions requires a deep understanding of the underlying semantics, and is expected to be performed by a relatively small number of application developers.

3. OVERVIEW OF THE SPRING EXAMPLE

The dual-spring design example is a case study specified by The US Air Force Materials Directorate to demonstrate an end-to-end Rosetta-based design activity. The case study requires design of a system comprising two coaxial wire springs, one shorter than the other. The inner spring must fit over a 1.0" diameter tube and must fit inside the outer spring. The outer spring must fit inside a 2.5" diameter tube. The maximum uncompressed length of the longer spring is 12.0", and the system must allow deflection of 6.0". For the first 4.0" of deflection, the system must exhibit a rate (force per unit deflection) of 50 lb/in, and for deflection between 4.0" and 6.0", the system must exhibit a rate of 75lb/in.

In addition to these problem-related requirements and constraints, equations are provided that relate the performance properties of a spring (rate and free length) to its physical properties (wire diameter, coil diameter and pitch, number of coils). Further constraints are specified for a spring to ensure that it does not assume a permanent deformation due to excessive compression. These constraints are expressed as inequalities in terms of the materials properties of the spring wire, including its modulus of rigidity, shear stress and shear yield stress.

4.   THE ROSETTA SPRING MODEL

The Rosetta model for the design example is comprised of two parts, presented in full in the Appendix to this paper. The first part is a parameterised model of a single spring, expressed as a Rosetta facet in the logic domain. We use this domain since we are dealing with a static model, not a dynamic model.

The formal parameters to the facet describe the physical characteristics of the spring: the wire diameter (d), the coil inner diameter (ID) and outer diameter (OD), the Modulus of Rigidity of the wire (G), the number of coils (N), and the coil pitch (p). Within the facet, a number of internal items are defined, representing the derived quantities mentioned in the spring equations. These include the spring rate (R) and desired free length (L) used in determining the performance of the spring, and the material properties used in the constraints.

The spring facet models a spring in terms of equations and inequalities that relate these parameters and derived quantities. We use the convention of labelling constraint terms with labels C1, C2, etc., formula terms with labels F1, F2, etc., and assertion terms with labels A1, A2, etc. Term C1 simply relates the inner and outer diameters to the wire diameter. C2 and C3 are constraints on the spring constant and number of coils determined from experience and given in the problem statement. F1 is given in the problem statement as a formula relating several variables, and is simply translated into Rosetta using predefined mathematical operators. Formulas represented in F2 and F3 are likewise given in the problem statement. N is the total number of coils. This, multiplied by the wire diameter, gives the height of the full compressed spring. F4 gives the minimum free (uncompressed) length of the spring (Lm): the fully compressed height (H) plus the maximum compression that will be applied. However, since the spring rate is non-linear over the last 20% of compression, the desired free length (L) is set to be longer than the minimum free length. This is expressed in F5. The next term, F6, expresses the coil pitch (p) in terms of the free length, wire diameter and number of active coils. C4 is a constraint that ensures that the spring is "spring-like" rather than being a column of wire.

The remaining terms in the spring facet define the corrected shear force applied to the spring on full deflection (Sk) and the shear yield strength of the wire (Sy). These terms are transliterations of the formulas provided in the problem statement. Finally C5 is the constraint that prevents the applied shear force exceeding the yield strength. If this constraint is violated, the spring "takes on a set," that is, it suffers a permanent deformation.

The second part of the Rosetta model for the design example is a facet that describes the composite system. The spring_system facet declares a number of variables representing properties of the inner and outer springs, as well as the overall system deflection and applied force. The facet also includes two instances of the single-spring facet, and a collection of equations and constraints that parallel the informal specification provided in the problem statement.

In particular, the facet includes equations that relate the overall system deflection to the deflections of the individual springs, and the overall system rate to the rates of the individual springs. The Modulus of Rigidity of the springs is given in the problem specification, but the other physical characteristics are required to be determined from the required overall rate and the physical constraints. Hence the actual parameters for the spring instances, describing the physical characteristics, are free variables. Their values depend on the rate for each of the springs, determined from the overall rate, and the various constraints that apply to the individual springs and to the composite system.

The apparent complexity of the formulas describing deflection of the system and the individual springs comes about from the fact that the problem does not specify which of the two springs is the longer. For the system rate (the sum of the two spring rates) to change at a deflection of 4", one spring must be 4" longer than the other. The longer spring must have a rate of 50lb/in, and the shorter, engaged when the deflection reaches 4", must have a rate of an additional 25lb/in. In principle, this could be achieved with either spring being the longer. Care was taken in constructing the formula not to bias the solution. If only one alternative is feasible, that fact should be a consequence of the constraints upon the system rather than an *a priori* statement in the model.

The formula F6 and the assertion A1 are not strictly required in the model. However, they are derived from statements in the problem definition that serve as clarifications of the rate specifications. They can be seen as "sanity checks," and so were included in the model as assertions for this purpose.

5. EVALUATION OF THE MODEL

The dual-spring system case study is one of the first modelling problems specified externally to the language development team to be attempted as part of the language validation process. It was undertaken largely by the first author after briefly reviewing the preliminary documentation on the language. Most of the effort in developing the model was spent in understanding the English-language and equation-based informal specification. Thereafter, expression in Rosetta was relatively straightforward. The major difficulties lay in determining which properties of the springs should be specified as parameters and which as exported properties. This question is still not clear, and we expect it will be treated as a matter of modelling style.

An important benefit of the declarative nature of the Rosetta description became evident during development of the model. The spring facet was developed from the perspective of starting with the physical parameters of the spring (e.g., wire diameter, coil diameter, etc.) and deriving the behavioral properties (rate, free length). However, when the facet was instantiated in a system model, the behavioral properties were given and the physical properties were to be determined. The declarative nature of the equations makes it possible to "drive the model backwards" in this way. Were the model expressed in an operational style, with variables being assigned from inputs, such an approach would be much more difficult.

In order to validate the Rosetta model, the first author prepared an Excel spreadsheet that encapsulated the formulas in the model. In fact, two spreadsheets were developed, one structured with the inner spring being the longer, and the other structured with the outer spring being the longer. This was done to make the Excel model manageable and solvable. The author attempted to solve the dual spring problem in each variation by using the solver facility in Excel. This facility allows the user to specify goals on dependent variables, constraints on independent variables, and to find values of independent variables that imply the goals. The author was not able to find a feasible solution to the dual-spring problem as specified. It remains unclear whether this is because the problem really is overconstrained, or whether the Excel model is too simplistic to allow identification of a feasible solution.

In addition to the Excel translation, an interface between Rosetta and the MATLAB environment has been developed. The MATLAB translation system takes parsed output from the standard Rosetta parser and generates equations suitable for evaluation in the MATLAB environment. An evaluation script was developed to evaluate the model over it operational environment. The model resulting from the automatic transformation evaluated favourably with respect to the hand generated Excel result.

6. CONCLUSION

The design case study described in this paper demonstrates the effectiveness of Rosetta as a specification and constraint language for describing physical systems such as the dual-spring system. In principle, the formal semantic basis of the language enables automatic tool-based analysis, such as checking for inconsistency, and automatic solution of equations.

While the power of Rosetta has been demonstrated in the domain of continuous, static systems, it is by no means limited to this domain. Other predefined domains allow description of the dynamic behavior of system, including continuous-time, discrete-time and state-based models. It is expected that further case studies will demonstrate the use of the language for requirements and constraint specification in these domains. However, one of the most exciting capabilities will be the expression, with a formal semantic basis, of the interactions between these domains. This remains one of the most significant contributions of the Rosetta language development effort.

REFERENCES

[1] Alexander, P,. C. Kong, and D. Barton, "The Rosetta Usage Guide," University of Kansas Technical Report available at http://www.ittc.ukans.edu/Projects/rosetta/
[2] Alexander, P., C. Kong, and D. Barton, "The Rosetta Functional Requirements Specification Domains," *Proceedings of the Hardware Description Language Conference (HDLCON'00)*, March 2000, Los Angeles, CA.
[3] Alexander, P., R. Kamath, D. Barton, "System Specification in Rosetta," *IEEE Engineering of Computer Based Systems Workshop and Symposium*, April 2000, Edinburgh, UK.
[4] The VHDL International Systems Level Design Language Committee Web page, http://www.intermetrics.com/sldl

APPENDIX: THE ROSETTA SPRING MODEL

```
facet spring ( d :: real;        // wire diameter
                OD :: real;       // outside diameter of spring
                ID :: real;       // inside diameter of spring
                G :: real;        // Modulus of Rigidity of wire
                N :: real;        // total number of coils
                p :: real         // pitch
              ) is
    export R, L;

    pi :: real is 3.1415926435898;

    D :: real is (OD + ID) / 2;  // mean diameter of spring
    C :: real is D / d;           // spring constant
    n :: real;                    // number of active coils
    H :: real;                    // fully compressed height
    R :: real;                    // Force/length (lb/inch)
    P :: real;                    // force to fully deflect spring
    l :: real;                    // maximum spring displacement
    Lm :: real;                   // minimum free length
    L :: real;                    // desired free length
    Tu :: real;                   // ultimate tensile strength of wire
    Ty :: real;                   // tensile yield strength
    Sy :: real;                   // shear yield strength
    S :: real;                    // uncorrected shear stress
    k :: real;                    // correction factor
    Sk :: real;                   // corrected shear factor
begin logic
    C1: (OD - ID) / 2 = d;   // dependencies between diameter parameters
    C2: C >= 4 and C =< 20; // empirical constraints on spring constant
    F1: n = (G * d^4) / (8 * R * D^3);
    C3: n >= 3;                   // empirical constraint on n
    F2: N = n + 2;                // for closed and ground spring
    F3: H = d * N;
    F4: Lm = H + l;
    F5: L = H + (l / 0.8);
    F6: p = (L - 2*d) / n;
    C4: p =< D;                   // pitch can't exceed coil diameter
    F7: Tu = 200000 * D^(-0.14);
    F8: Ty = 0.75 * Tu;
    F9: Sy = Ty * 0.577;
    F10: P = R * l;
    F11: S = (8 * P * D) / (pi * d^3);
    F12: k = (4*C - 1) / (4*C - 4) + 0.615 / C;
    F13: Sk = S * k;
    C5: Sk < Sy;                  // shear force < wire shear yield strength
end spring;
// ----------------------------------------------------------------
```

```
facet spring_system is

    outer_d :: real;   outer_OD :: real;   outer_ID :: real;
    outer_G :: real is 11.5E6;   // music wire
    outer_N :: real;   outer_p :: real;
    inner_d :: real;   inner_OD :: real;   inner_ID :: real;
    inner_G :: real is 11.5E6;   // music wire
    inner_N :: real;   inner_p :: real;

    system_L :: real;                // system free length

    system_R :: real;                // system rate
    inner_R :: real;   outer_R :: real;

    system_deflection :: real;
    inner_deflection :: real;   outer_deflection :: real;

    force :: real;

begin logic

    outer_spring: spring( outer_d, outer_OD, outer_ID,
                          outer_G, outer_N, outer_p);

    inner_spring: spring( inner_d, inner_OD, inner_ID,
                          inner_G, inner_N, inner_p);

    C1: outer_OD < 2.5;       // outer spring must fit inside 2.5" tube
    C2: outer_ID > inner_OD; // outer spring must fit over inner spring
    C3: inner_ID > 1.0;       // inner spring must fit over 1.0" tube

    F1: system_L = max(outer_spring.L, inner_spring.L);
    C4: system_L < 12.0;

    C5: system_deflection >= 0.0 and system_deflection =< 6.0;

    F1: inner_deflection =
            if system_deflection < (system_L - inner_spring.L) then 0
            else system_deflection - (system_L - inner_spring.L);
    F2: outer_deflection =
            if system_deflection < (system_L - outer_spring.L) then 0
            else system_deflection - (system_L - outer_spring.L);

    F3: inner_R =
            if inner_deflection = 0 then 0 else inner_spring.R;
    F4: outer_R =
            if outer_deflection = 0 then 0 else outer_spring.R;

    F5: system_R = outer_R + inner_R;

    C6: system_R =
            if system_deflection <= 4.0 then 50 else 75;

    F6: force = system_R * system_deflection;

    A1: (system_deflection = 0 => force = 0)
        and (system_deflection = 4.0 => force = 200)
        and (system_deflection = 6.0 => force = 350);

end spring_system;
```

Transformational System Design Based on Formal Computational Model and Skeletons

Wenbiao Wu, Ingo Sander, Axel Jantsch

Royal Institute of Technology, Stockholm, Sweden

Key words: Formal System Design, Program Transformation, Haskell, Skeleton

Abstract: The Formal System Design methodology *ForSyDe* [1,2,3] is extended by a systematic refinement methodology based on transformations, which gradually transforms a high-level, function oriented system description into a synthesizable model. We group transformations according to three different criteria: (1) whether they preserve semantics or they constitute a design decision; (2) whether they are simple rewriting rules or complex patterns; (3) whether they transform combinatorial functions, skeletons or data types. We illustrate the use of transformations with three examples taken from an ATM based network terminal system.

1. INTRODUCTION

System level functional validation has been identified as one of the most severe obstacles to increased design productivity and quality. It is most likely that this will not fundamentally change unless a rich portfolio of validation techniques becomes an integral part of the system specification and design process. In addition to simulation, which is the main validation vehicle today, formal analysis and verification techniques must be available

P.J. Ashenden et al. (eds.), System-on-Chip Methodologies & Design Languages, 171–185.
© 2001 *Kluwer Academic Publishers. Printed in the Netherlands.*

to the designer. Unfortunately, the full potential of formal techniques can only be exploited when they become an equal partner to simulation and design activities, which means that modeling and design techniques have to adapt and change significantly. Most current and proposed design languages such as VHDL, C++, Superlog [4], SystemC [5], etc., are foremost simulation languages. Considerations about formal analysis and verification are only taken up as an afterthought. This will always be insufficient because the semantics of these languages fundamentally limits the kind of formal analysis that can be performed.

We have proposed *ForSyDe* (Formal System Design) [1,2,3], a system modeling method based on a formal functional model and on skeletons with the objective to eventually develop both powerful formal analysis techniques and a full fledged refinement and synthesis process. In [1] we have argued that a carefully designed modeling method is the solid basis for efficient verification and refinement. Now we take the next step and propose a refinement technique, which takes formal verification into account up front. It is based on transformations, which allows to split the system validation into numerous small steps, one separate verification step for each applied transformation. Thus, the seemingly impossible task of verifying the functionality of an entire complex system becomes feasible because it is reduced to many smaller and simpler steps.

The main contribution of this paper is the extension of *ForSyDe* methodology with a transformational refinement process of the system model. We outline different types of transformations, which together are powerful enough to allow refinement of a system level functional specification into a synthesizable description. The technique is illustrated by applying selected transformations on a system model of an ATM switch.

We review the related work in this area in section 2, introduce our system model in section 3, discuss transformational design in section 4, illustrate transformation examples in section 5 and conclude our paper in section 6.

2.   RELATED WORK

Transformational system design supports the idea of designing at a higher level of abstraction which implies describing the behavior of the required system in high level languages and then transforming this into a description at a lower level, possibly in a hardware description language. Such kind of design methodology can reduce the overall design time and ensure the correctness of the implemented system. A great deal of the pioneering work in transformation systems was undertaken by Burstall and Darlington [6]. Their system transformed applicative recursive programs to imperative ones and their ideas have heavily influenced today's transformation systems.

Haskell [7] is a modern functional language that provides a common ground for research into functional languages and functional programming. However, now it has also been used in hardware design. In the Hawk project [8], Hawk is a library of Haskell functions that are appropriate building blocks (or structural units) for describing the micro-architecture level of a microprocessor. In their work, Hawk has been used to specify and simulate the integer part of a pipelined DLX microprocessor. Lava [9] is a collection of Haskell modules. It assists circuit designers in specifying, designing, verifying and implementing hardware. O'Donnell [10] has developed a Haskell library called Hydra that models gates at several levels of abstraction ranging from implementations of gates from using CMOS and NMOS, up to abstract gate representations using lists to denote time-varying values. Möller [11] provided a deductive hardware design approach for basic combinational and sequential circuits. The goal of his approach is the systematic construction of a system implementation starting from its behavior specification. Gofer/Haskell (Gofer is an interpreter for Haskell) has been used in his work to formulate a functional model of directional, synchronous and deterministic systems with discrete time. However, all of the above approaches are targeted on gate-level or circuit designs which differ significantly from our approach. Reekie [12] used Haskell to model digital signal processing applications. He modeled streams as infinite lists and used higher-order functions to operate on them. Finally, correctness-preserving transformations were applied to transform a system model into a more effective representation. However, in Reekie's work, the target applications are data flow networks and the final representation is not synthesized. In our approach, we targeted at both data flow and control intensive applications and the final system is synthesized to a VHDL and C model.

There are also some other languages to support hardware design. The Ruby language, created by Jones and Sheeran [13], is a circuit specification and simulation language based on relations, rather than functions. The target applications are regular, data flow intensive algorithms, and much of its emphasis is on layout issues. In contrast, our approach is based on a functional language, addresses data flow and control dominated applications, uses a fully fledged functional language, and links to commercial logic synthesis tools rather than dealing with layout directly. HML [14] is a hardware modeling language based on the functional language ML, which is a functional language similar to Haskell used in our approach. However, HML attempts to replace VHDL or Verilog as hardware description languages, while we propose a hardware and system specification concept on a significantly higher abstraction level with a very different computational model.

3.   SYSTEM MODEL

Today, system design often starts with a description of concurrent processes communicating with each other by means of complicated mechanisms (shared variable, remote procedure call or asynchronous message passing). We have argued [15] that such a system model does not serve as a good starting point for system design, because many design decisions are already taken. In particular, it is difficult to change the process structure of the system model, i.e. to move functions from one process to another, since this will require the redesign of the communication structure between the processes.

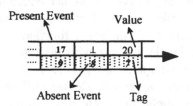

Figure 1. A signal is a set of events

We have addressed this problem in our design methodology by adopting a purely functional specification of the system based on data dependencies and written in the functional language Haskell, for which a formal semantics exists. Our computational model is based on the synchronous hypothesis which assumes the outputs of a system are synchronized with the system inputs, while the reaction of the system takes no observable time. For a formal definition of the computational model, we use the denotational framework of Lee and Sangiovanni-Vincentelli [16]. It defines a signal as a set of events, where an event has a tag and a value. A signal is shown in Figure 1. Here events are totally ordered by their tags. Events with the same tag are processed synchronously. A special value '⊥' ("bottom") is used to model the absence of an event. Absent events are necessary to establish a total ordering among events for real time systems with variable event rates.

Compared with finite state machines which are more operational in their nature, this approach leads to a computational model with a definite denotational flavor.

The system is modeled by means of concurrent processes which are supported by the use of skeletons. Skeletons are higher-order functions whose inputs and outputs are signals. Elementary functions are the basic components inside the skeleton. In terms of a clocked system, elementary functions can model the behavior of system components in one clock cycle. On the other hand, each skeleton has a representation in the design library and can be directly mapped into hardware or software component based on the information from a design library.

Due to its formal nature of the system model, which can be regarded as consisting of mathematical entities, it can be further analyzed by using mathematical methods, which in turn can also be mechanized. For example, the formal model can be integrated with theorem provers, model checkers and other formal analysis and verification techniques, which can in some way guarantee the correctness of the specification model.

4. TRANSFORMATIONAL DESIGN

4.1 Introduction

With transformational design, we mean starting the system design with a formal specification model and then applying transformation techniques on this model to get a synthesizable system model, i.e. the specification can be transformed through a series of steps to a final synthesizable system model which meets the system requirement and constraints imposed by designers (Figure 2). These steps can be either semantics preserving or non-semantics preserving. The later introduces design decisions imposed by the designer. The transformation result of each step is sufficiently close to the previous expression that the effort of verifying the transformation is not excessive. The benefits of this approach are:

- Each step in the system refinement is closer to the final implementation.
- The final implementation is either a true implementation of the initial specification which is guaranteed by the use of semantics preserving transformation rules in the library, or can be proofed correct for

certain assumptions w.r.t non-semantics preserving transformations which are introduced by design decisions.

- The design process and design decisions are documented, thus the whole process is repeatable.
- A transformation approach made up of a sequence of small steps is very effective and much more efficient than theorem proving which is usually very difficult because the gap between implementation and specification is much larger than the gaps between each transformation steps.

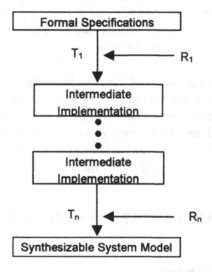

Figure 2. Transformational system design

Of course, there are still difficulties in applying a purely transformational approach to large system design. However, the incorporation of this approach into the design methodology will offer the opportunities to improve the design process. For example, in HW/SW co-design, transformation techniques can be employed in the system's custom parts that are not covered by pre-designed building blocks (IPs).

4.2 Refinement of the system model by transformation

We distinguish two levels of transformations:

- *Transformation rules* are primitive transformations as developed over the last three decades [6,17]. They include
  - Definition: Introduce new functions to the system model;
  - Unfolding: Replace a function call with its definition;

- ♦ Folding: the opposite of unfolding;
- ♦ Abstraction: Abstract common expressions used in system models into one separate function to be reused;
- ♦ Algebraic rules;
- **Transformation patterns** describe more complex transformations which in many cases can be application and implementation related. For instance, if a finite state machine model contains a large and regular structured state, a transformation pattern can transform this into a simpler finite state machine and a memory skeleton. In the synthesis step such a model can later be implemented as a memory with memory controller, which is far more suitable than an FSM-implementation with many registers. Essentially, transformation patterns capture intelligent design techniques.

Orthogonal to this we classify transformations in the following way:

- **Semantic preserving transformations** do not change the functionality of the system, but they may have an impact on non-functional properties such as performance and cost. Semantic preserving transformations have the beautiful property, that they can be freely applied without changing the functionality. This greatly alleviates the validation and allows for fast design space exploration. Transformation rules are always semantic preserving, but transformation patterns may or may not be.
- **Decision transformations** inject a design decision into the refinement process. During the refinement process from a high abstraction level down to the implementation a number of decisions must be taken which also change the functionality. For instance, the high-level functional model will in many cases use infinite buffers (FIFO) for communication between entities. However, during refinement the infinite buffers must be reduced to finite buffers which invariably changes the behavior, because the finite buffers may overflow for certain input sequences. Decision transformations may be proved by adding additional assumptions. E.g. if we assume that the input sequences are constrained in such a way, that the finite buffers will never overflow, we can proof that the two descriptions behave identically. This has the additional benefit of making all these assumptions explicit and part of the system specification.

In addition, we can group the transformations in the following way: skeleton-based transformations on skeletons, local transformations on elementary functions which are usually the essential part of a skeleton, and data-type transformations for data types. This is also orthogonal to other classifications.

- *Skeleton-based transformations* are done on a higher level than local transformations on elementary functions. They are usually employed on the architecture model and will depend heavily on the design library which contains implementations for skeletons and library elements. In our design methodology, this kind of transformation will include merge and split skeletons, move functions between skeletons to minimize the communications between skeletons etc. and share elementary functions between skeletons.
- *Local transformations* are employed on the elementary functions inside the skeleton. They can be introduced by skeleton-based transformations, for example merging and splitting skeletons usually give rise to the reconstruction of internal elementary functions which permits further optimization.
- *Data type transformations* are usually employed to transform one list to multi-lists or vice versa, split a signal into several signals to be processed by several processes simultaneously to acquire high efficiency, transform data structures (e.g. trees and lists) to meet the specialized system implementation and transform unbounded data types to bounded data types.

During design exploration, these transformations are employed on the system model to optimize the system and thereby to meet the specified constraints. The system model is successively refined by means of applying small transformational steps to the model in order to receive a synthesizable system model, which meets the specified constraints. The refinement process is also supported by estimation values for possible transformations.

4.3 Synthesis of the synthesizable system model

After the refinement process the HW parts of the synthesizable system model are synthesized into VHDL, while the SW parts are synthesized into C. Here we use the HW and SW interpretations of the skeletons. The VHDL model is then further processed by logic synthesis tools while the C-code is compiled for a specific processor [2,3].

5. EXPERIMENTS IN THE ATM CASE STUDY

In this section we illustrate the transformational design methodology by several examples from our case study, an ATM switch. A more detailed description of the design of this ATM switch can be found in [1]. The ATM switch has operation and maintenance functionality. This case study has the following aspects:

- The original specification model is written completely in Haskell with skeletons used.
- The aforementioned transformation rules and patterns are employed manually during the design explorations.
- The design exploration utilizes the results from library estimation as one of the exploration criteria.
- The final system is a functional model which can be easily synthesized into a VHDL and C model.

Example I

First we introduce the skeletons used in these examples:

mapT is the skeleton that applies a function *f* on to the values of all events in a signal. It is based on the function *map* which maps a function on a list. It is defined like this in the functional specification:

mapT :: (a -> b) -> TimedSignal a -> TimedSignal b
mapT f (Sig xs) = Sig (map f xs)

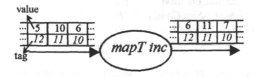

Figure 3. Skeleton *mapT*

Figure 3 shows skeleton *mapT* with elementary function *inc* which increments the value in each event.

filterT is the skeleton that filters out the events in a signal which doesn't satisfy property *p*. Here is its definition:

filterT :: (a -> Bool) -> TimedSignal a -> TimedSignal a
filterT p = mapT (check p)
  where check p x = if p x then x else Absent

Figure 4. Skeleton *filterT*

Figure 4 shows the skeleton *filterT* with elementary function *even* which filters out events with odd values.

In the ATM system, we use the following functional description which checks the status of a virtual path and generates OAM cells if a certain condition is met. This is also shown in the first part of Figure 5.

sendDownstream = mapT action.filterT isSendCondition
   where
      isSendCondition (Present (vpi, (VPI_AIS, 1000)))   = *True*
      isSendCondition (Present (vpi, (_, _)))   = *False*

action (Present (vpi, (VPI_AIS, time))) = Present (F4_OAM (VPI vpi) VPI_RDI)
action _   = *Absent*

sendDownStream first filters out the events in incoming signal according to *isSendCondtion*, then it takes *action* on the outgoing signal from the previous process.

This specification can be transformed to:

newSendDownstream
  = *mapT action . filterT isSendCondition*   *(0)*
  = *mapT action . mapT (check isSendCondition)(1)*
  = *mapT (action . (check isSendCondition))*   *(2)*
  = *mapT newAction*   *(3)*

newAction Present (vpi, (VPI_AIS, 1000)) = *Present (F4_OAM (VPI vpi) VPI_RDI)*
newAction _   = *Absent*

Step (0) is the definition of *newSendDownStream*. The transformation step from (1) to (2) is a skeleton-based transformation which is based on the rule *mapT f . mapT g = mapT (f.g)*. However, the transformation step from (2) to (3) is an elementary-based transformation, which is aimed to optimize the elementary function inside the skeleton. This example also shows that transformation allows the optimization of internal functions. The advantage of this transformation is apparent if we compare them in Figure 5. The original structure will usually be implemented in two processes, either HW or SW. There is intensive communication between these two processes. However, in the new specification, only one process is needed in the final implementation.

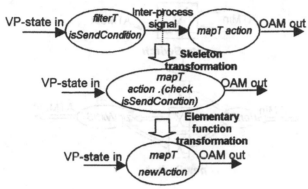

Figure 5. Transformation on *sendDownstream*

Example II

To get the system work reliably under some critical environments, we need some fault tolerance in the system design. In the ATM switch we can duplicate the switch to achieve this goal. This is shown in Figure 6, where *atmSwitch* is the original switch module and *newSwitch* is the resulting system which is derived from *atmswitch*. The definition of *newSwitch* is as following:

```
newSwitch atmin = (atmOut1, atmOut2)
  where
    atmOut1 =  zipwith3T selectF atmUpOut1 atmMiddleOut1 atmDownOut1
    atmOut2 =  zipwith3T selectF atmUpOut2 atmMiddleOut2 atmDownOut2
    (atmUpOut1, atmUpOut2)            = atmSwitch atmUpin
    (atmMiddleOut1, atmMiddleOut2)    = atmSwitch atmMiddlein
    (atmDownOut1, atmDownOut2)        = atmSwitch atmDownin
    (atmUpIn, atmMiddleIn,atmDownIn)  = fan3T id id id atmin

fan3T f g h x = (f x, g x, h x)
zipWith3T :: (a -> b -> c) -> TimedSignal a -> TimedSignal b -> TimedSignal c
zipWith3T f (Sig (x:xs)) (Sig (y:ys)) (Sig (z:zs)) = Sig (f x y z: zipWith3 f xs ys zs)
```

Function *fan3T* fans one ATM stream into three which are then fed into three ATM switches in *newSwitch*. The results of all switches are later sent to (*zipWith3T selectF*) which selects one value as output. Function *selectF* can be defined in different ways. A common implementation can be majority vote, which compares the inputs and selects the value, which is equal in two or three of the inputs. The correctness of this transformation can be easily proved. As a result, the object system is more reliable than the original one. Similar transformations to make the design more parallel can be beneficial also for performance reasons.

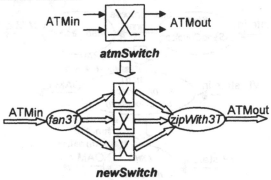

Figure 6. Transform one module to three

Example III

The previous examples are semantics preserving transformation. There are also non-semantics preserving transformations. We will illustrate this with the FIFO design in this ATM case study.

In the ATM example, an unconstrained FIFO is defined as:

unconstrainedFifoT :: TimedSignal [a] -> TimedSignal a
unconstrainedFifoT = mooreS fifoState fifoOutput []

The function *fifoState* is used to calculate the new state of the buffer. The function *fifoOutput* analyses the buffer and outputs the first element.

In the final implementation, the FIFO will only have a fixed buffer size and can only consume a fixed number of items during one event cycle. To solve this, we obtain a template by replacing the function *fifoState* with a new function *constrainedFifoState* with two additional parameters b for the buffer size and i for the number of parallel inputs.

constrainedFifoTemplate :: Int -> Int -> TimedSignal [a] -> TimedSignal a
constrainedFifoTemplate b i = mooreS (constrainedFifoState b i) fifoOutput []

We can then build instances of constrained FIFOs by specifying the parameters b and i. For example: (Figure 7)

constrainedFifoT_b8_i4 :: TimedSignal [a] -> TimedSignal a
constrainedFifoT_b8_i4 = constrainedFifoTemplate 8 4

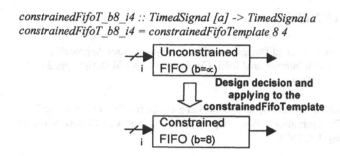

Figure 7. Decision transformations on FIFO

The result of this transformation is much closer to the final synthesizable model and this transformation can be used in several parts of the case study.

We have illustrated some transformation techniques with the ATM case study. The purpose of this case study is to validate our transformational system design methodology. Furthermore, through this case study, we have obtained a better understanding of the required transformations and how they can be mechanized to solve real industry problems. However, this is only a small subset of the complete transformation library which will eventually include a rich set of transformation patterns. Besides the transformations in this case study, it also includes memory-based transformations for FSM, communication transformations and other transformations on skeletons. This set of transformation patterns combined with the various transformation rules should be sufficient to derive a synthesizable model from the functional specification.

6.   CONCLUSION AND FUTURE WORK

We introduced transformational refinement to our design methodology for HW/SW co-design which starts with a formally defined functional system model and abstracts from the implementation details. The system model is stepwise refined by formally defined transformations leading to a synthesizable system model. Due to the use of skeletons the synthesizable model can be synthesized into an efficient implementation incorporating HW and SW parts.

As the next step of our current work, we will focus on how to mechanize some of the transformations we have proposed and how to verify these transformations.

184

7. REFERENCES

[1] I. Sander and A. Jantsch, "Formal Design Based on the Synchronous Approach", Proceedings of the Twelfth International IEEE Conference on VLSI Design, pp. 318-324, Goa, India, January 7-9, 1999.

[2] I. Sander and A. Jantsch, "System Synthesis Utilizing a Layered Functional Model", Proceedings of the 7th International Workshop on Hardware/Software Codesign, pp. 136-141, Rome, Italy, May 3-5, 1999.

[3] I. Sander and A. Jantsch "System Synthesis Based on a Formal Computational Model and Skeletons", Proceedings of IEEE Workshop on VLSI'99 (WVLSI'99), pp. 32-39, Orlando, Florida, USA, April 8-9, 1999.

[4] P. L. Flake and Simon J. Davidmann, "Superlog, a unified design language for system-on-chip", Proceedings of the Asia and South Pacific Design Automation Conference (ASP-DAC), pp. 583 -586, 2000.

[5] S. Y. Liao, "Towards a New Standard for System-Level Design", Proceedings of the Eighth International Workshop on Hardware/Software Codesign (CODES), pp. 2-6, May 2000.

[6] R. M. Burstall and J. Darlington. A Transformation System for Developing Recursive Programs. Journal of the ACM, 24(1), pp. 44-67, January 1977.

[7] P. Hudak, J. Peterson and J. H. Fasel, A Gentle Introduction to Haskell. At http://www.haskell.org/tutorial.

[8] J. Mattews, B. Cook and J. Launchbury, Microprocessor Specification in Hawk, In Proceedings of IEEE International Conference on Computer Languages, Chicago, May 14-16, 1998.

[9] P. Bjesse, K. Claessen, M. Sheeran and S. Singh, Lava: Hardware Design in Haskell. In ACM International Conference on Functional Programming, Baltimore, Maryland, 1998.

[10] J. O'Donnell, From Transistors to Computer Architecture: Teaching Functional Circuit Specification in Hydra, In Symposium on Functional Programming Languages in Education, July 1995.

[11] B. Möller, Deductive Hardware Design: A Functional Approach, Report 1997-09, Institute of Computer Science at the University of Augsburg, 1997.

[12] H. J. Reekie, Realtime Signal Processing, PhD Thesis, University of Technology at Sydney, Australia, 1995.

[13] G. Jones and M. Sheeran, "Circuit Design in Ruby", in Formal Methods for VLSI Design, North Holland, edited by J. Staunstrup, 1990.

[14] Y. Li and M. Leeser, "HML: A Novel Hardware Description Language and Its Translation to VHDL", IEEE Transactions on VLSI Systems, Vol 8(1), pp. 1-8, Feb. 2000.

[15] A. Jantsch and I. Sander, "On the Roles of Functions and Objects in System Specification", in Proceedings of the International Workshop on Hardware/Software Codesign, 2000.

[16] E. A. Lee and A. Sangiovanni-Vincentelli, "A Framework for Comparing Models of Computation," IEEE Transactions on CAD, Vol. 17, No. 12, December 1998.

[17] A. Pettorossi and M. Proietti, "Rules and Strategies for Transforming Functional and Logic Programs", ACM Computing Surveys, Vol. 28, No. 2, pp. 360-414, June 1996.

Models of asynchronous computation

Mark B. Josephs
Centre for Concurrent Systems and Very Large Scale Integration, School of CISM, South Bank University, London, UK

Key words: trace-theoretic models; asynchronous processes; receptive processes; loosely-coupled processes; Brock-Ackermann anomaly

Abstract:

Synchronous and asynchronous computational paradigms are contrasted. The traces, failures and divergences models of asynchronous processes illustrate C.A.R.Hoare's formal approach to the modelling of such paradigms. (The failures model expresses nondeterminism.) A simplified form of failures model is shown to be suitable for receptive processes. These compute outputs in response to inputs and can be either tightly synchronised or loosely coupled. The failures model is used to clarify the reason why an example devised by J.D. Brock and W.B. Ackermann is problematic for a naïve generalisation of G. Kahn's history functions.

1.   SYNCHRONOUS VERSUS ASYNCHRONOUS

There are a variety of computational paradigms that are concerned with networks or circuits. These paradigms are applicable to the design of systems in which the computational process is performed by software, by hardware or by a combination of the two.

Often synchronous (clocked) computation is contrasted with asynchronous (self-timed) computation. The former involves all processes operating in lockstep, whereas the latter involves each process operating at its own rate. Synchronous computation is attractive because processes are

187

P.J. Ashenden et al. (eds.), System-on-Chip Methodologies & Design Languages, 187–192.
© 2001 *Kluwer Academic Publishers. Printed in the Netherlands.*

never blocked. On the other hand, much of the time a process may not be doing anything useful. Asynchronous computation is harder to understand because processes can become blocked waiting for each other. There is also the highly undesirable possibility of a network becoming deadlocked, namely, when all its processes are blocked.

Without a clock to regulate them, asynchronous processes can also become unstable (divergent), e.g., by engaging in infinite chatter or by oscillating in some undesirable way. The instability of one process risks destabilising the network as a whole.

2.   TRACE-THEORETIC MODELS OF ASYNCHRONOUS PROCESSES

According to Hoare [Hoa90] "A model of a computational paradigm starts with choice of a carrier set of potential direct or indirect observations that can be made of a computational process. A particular process is modelled as the subset of observations to which it can give rise."

The carrier set Σ^* for asynchronous computation is often taken to consist of finite sequences ("traces") of "events" belonging to an alphabet Σ. Then a model T of an asynchronous process would be a subset of Σ^*. For example, given an alphabet of events that includes `dial.`n, for n a four-digit number, `connect` and `disconnect`, the sequences < `dial.7413, connect, disconnect, dial.6262` > and < `dial.7413, disconnect` > are both in the carrier set. A model of a modem might include the former, but not the latter.

Assuming that all that is signified by membership of T is that the events in a trace can occur in that order, then one would expect the healthiness conditions

$$<> \in T \tag{1}$$

$$s \leq t \wedge t \in T \Rightarrow s \in T \tag{2}$$

to be satisfied by T, i.e., it contains the empty sequence and is prefix-closed [Hoa85]. Also, the deadlocked process (which is unable to engage in any further events) is modelled by { <> }.

Hoare [Hoa90] also explains that "A product meets a specification if its potential observations form a subset of its permitted observations." In particular, the relationship

$$T_P \subseteq T_S \qquad\qquad\qquad\qquad (3)$$

must hold between the traces model of a product P and that of its specification S. Thus, if one is only interested in the traces of a process, a deadlocked process meets every specification!

The blocking of a process when it fails to engage in any of a set of events that are offered to it can be captured by means of the carrier set $\Sigma^* \times 2^\Sigma$ [Hoa85]. With "failures" F it is possible to describe nondeterministic processes. For example, a modem might sometimes refuse to connect, $(< \texttt{dial.7413} >, \{ \texttt{connect} \}) \in F$, and sometimes succeed, $< \texttt{dial.7413}, \texttt{connect} > \in T$. A deadlocked process only meets those specifications for which $(<>, \Sigma) \in F_S$.

We do not consider the healthiness conditions for F here because in the next section we shall see that many asynchronous computational paradigms allow refusal sets to be simplified out of existence.

Finally, another model D with carrier set Σ^* can be interpreted as comprising (indirect) observations of a process diverging. It is usual to take a pessimistic view of divergence, placing no constraints on a divergent process:

$$s \in D \Rightarrow st \in D . \qquad\qquad\qquad\qquad (4)$$

3.   RECEPTIVE PROCESSES

So far we have not considered the causality of events. In many computational paradigms it is natural, however, to distinguish between input events and output events, so that Σ is partitioned into input alphabet I and output alphabet O [vSne85]. Having made this distinction, we can consider asynchronous computational paradigms in which processes must be receptive to input [Dil89]:

$$s \in T \wedge t \in I^* \Rightarrow st \in T . \qquad\qquad\qquad\qquad (5)$$

The idea is that a process cannot block input and in return is never blocked when it attempts to output. Note that if input arrives when it is not expected, a receptive process is allowed to diverge.

The only kind of failure that now can be (indirectly) observed is failure to output, which happens because the process is either quiescent or divergent [Jos92]. The carrier set for F is then simply Σ^* and a deadlocked process only meets those specifications for which $<> \in F_S$.

Traces, failures and divergences are related by

$$D \subseteq F \subseteq T. \tag{6}$$

Every trace is extensible by output events to a failure:

$$s \in T \Rightarrow \exists t \in O^*.\, st \in F. \tag{7}$$

Note that this treats a process that can output forever without requiring further input as unhealthy.

In order to ensure that our model of receptive processes is a complete partial order, we require bounded nondeterminism:

$$\{\, t \in O^* \mid st \in F \,\} \text{ is finite} \vee s \in D. \tag{8}$$

In [Jos92] we see how to use the weave operator [vSne85] to determine the parallel composition of receptive processes (assuming unbuffered communication channels) and to conceal output events simply by restricting observations. As a result, a network can be treated as a single process, allowing a hierarchical approach to be taken to system decomposition and verification.

4.   LOOSELY-COUPLED PROCESSES

Dataflow networks [Kah74], delay-insensitive circuits [Udd86] and handshake circuits [vBe93] are further asynchronous computational paradigms. In these cases, communications between sender and receiver are subjected to arbitrary delays. Nevertheless, our model for receptive processes of the previous section can be easily adapted to handle loosely-coupled processes. The idea, as in [Udd86], is to account for the effect of the wrapper created by the communication channels in presenting the set of potential observations of a process.

In particular, the sets of traces, failures and divergences are closed under reordering, e.g.,

$$sabt \in T \Rightarrow sbat \in T \tag{9}$$

where events a and b involve different channels, and $a \in O$ or $b \in I$ [Jos92].

5.   FAILURES VERSUS HISTORY FUNCTIONS

It is tempting to generalise from "history functions" [Kah74] (between input streams and output streams) to history relations in order to model nondeterminism. The problem with this approach can be seen as follows:

Given an input alphabet $\{a\}$ and an output alphabet $\{b\}$, consider first a process P that outputs twice unless it receives input before it has output anything, in which case it may output only once. Formally, $< b, b >$, $< b, a, b >$, $< a, b, b >$ and $< a, b >$ are failures, but $< b, a >$ is not.

Consider next a process Q that outputs twice unless it receives input before it has done so, in which case it may output only once. Thus, the failures of Q include those of P, but also include $< b, a >$. P and Q are distinct; in fact, P is more deterministic than Q.

Unfortunately, the distinction between P and Q is lost when their behaviour is characterised by the relationship between their input stream and their output stream. This relationship includes $< > \mapsto < b, b >$, $< a > \mapsto < b >$ and $< a > \mapsto < b, b >$.

This is the essence of the Brock-Ackermann anomaly [BA81].

A recent tutorial [LVS98] proposes "tagged signal" models of computation. The authors show how to express history functions for deterministic dataflow networks in their framework, but do not consider the nondeterministic case. There is a way of generalising history functions to avoid the Brock-Ackermann anomaly [SN85], so their framework may still be applicable. Nevertheless, the trace-theoretic models of asynchronous processes presented here are attractively simple.

ACKNOWLEDGEMENTS

The support of the UK Engineering and Physical Sciences Research Council under grant GR/M51567 and the European Commission under contract no. 21949 is acknowledged.

REFERENCES

[BA81] J.D. Brock, W.B. Ackermann. Scenarios: a model of non-determinate computation. Lecture Notes in Computer Science,107, 1981, pp. 252-259.

[Dil89] D.L. Dill. Trace theory for automatic hierarchical verification of speed-independent circuits. ACM Distinguished Dissertation Series, MIT Press, 1989.

[Hoa85] C.A.R. Hoare. Communicating Sequential Processes. Prentice-Hall International Series in Computer Science, 1985.

192

[Hoa90] C.A.R. Hoare. Let's Make Models (abstract). In: Proc. CONCUR'90, Lecture Notes in Computer Science, 458, 1990, p. 32.

[Jos92] M.B. Josephs. Receptive process theory. Acta Informatica, 29:17-31, 1992.

[Kah74] G. Kahn. The semantics of a simple language for parallel programming. In: Proc. Information Processing 74, North-Holland, 1974, pp. 471-475.

[LVS98] E.A. Lee, A. Sangiovanni-Vicentelli. A Framework for Comparing Models of Computation. IEEE Transactions on CAD, 17(12):1217-1229, 1998.

[SN85] J. Staples, V.N. Nguyen. A fixpoint semantics for nondeterministic data flow. Journal of the ACM, 32(2):411-444, 1985.

[Udd86] J.T. Udding. A formal model for defining and classifying delay-insensitive Circuits. Distributed Computing, 1(4):197-204, 1986.

[vBe93] K. van Berkel. Handshake Circuits: An asynchronous architecture for VLSI programming. Cambridge International Series on Parallel Computation, 5, Cambridge University Press, 1993.

[vSne85] J.L.A. van de Snepsheut. Trace Theory and VLSI Design. Lecture Notes in Computer Science, 200, 1985.

A mixed event-value based specification model for reactive systems

Norbert Fristacky, Jozef Kacerik, Tibor Bartos
Slovak University of Technology Bratislava, Ilkovicova 3, 811 09 Bratislava

Key words: digital system, specification, agent, event, finite state machine

Abstract: This paper deals with a behavioral specification model for reactive systems based on (behavior executing) agents and on processes that contain agents as their atoms and describe their compositions. An agent specifies a part of the system's input-output communication with its environment as a set of finite sequences of input and output values over a finite time interval in which all sequences contain discrete time stamps, and the final state at the end of the time interval. The time stamps are explicitly specified by timing events. Agents may contain a set of timing rules that specify the timing discipline which must hold in the continuous time interval in order to obtain the correct function. There is also a starting mechanism, by means of which the execution of agents and processes may be initialized.

1.   INTRODUCTION

At present, in designing digital devices, description languages such as VHDL or Verilog are commonly used. However, it is often more convenient to use some more abstract specification means that can be automatically translated to such a language. Reasons for this may rest in a better specification capture capability, and in making the development of specifications more systematic and easier to verify. Examples of such efforts are various graphically oriented specification means like Statemate[1] or SpecCharts[2], specification languages based on process algebras[17], finite state machines [9], etc.

In this paper we introduce a behavioral specification model for reactive systems which is based on formal (a given behavior executing) entities called   agents   and on processes that contain agents as their atoms and

P.J. Ashenden et al. (eds.), System-on-Chip Methodologies & Design Languages, 193–204.
© *2001 Kluwer Academic Publishers. Printed in the Netherlands.*

express compositions of agents executions in time (e.g. concatenation, parallel composition, etc.). An agent specifies (executes) partial deterministic system behavior over a finite time interval. Agents may contain a set of timing rules that specify the quantitative timing constraints which are defined over an explicitly defined discrete time structure. An agent describes:

- a part of the system's input-output communication with its environment, which is defined by a so-called communication set. It is a set of sequences of input and output values observed over discrete time points which are specified by chosen changes of variables, called timing events.
- the final state of the system at the end of the time interval which depends on the indicated input-output sequence and the initial state.

We define an algebra of agents as a basis for composition of agents, and to introduce processes that express such compositions. The semantics of a process is an agent, i.e. if we consider a process to be an implementation of a given agent, its correctness and consistency can be verified by checking the equivalence of two agents - the semantics of the process and the given specification. This introduces a hierarchy which can be employed in formal verification during the process of specification design.

We also introduce a starting mechanism as a means to initialize the execution of agents and processes by environmental events. It allows us to specify resetting and exception handling.

The descriptive text is interleaved with an example – fragments of the specification of the well-known DLX processor [18] (given on an higher level of abstraction in which no clock neither any DLX I-O communication protocol with its environment have been introduced yet). To specify DLX the formal specification language HSSL[0] is used. HSSL has been developed by the authors and is based on the described specification model.

2.   MOTIVATION

Let's suppose we have a digital system represented as a black box with finite number of inputs and outputs. If we have no information about the structure of the system, we have to describe its behavior strictly by some relations over the values observed on the systems' inputs and outputs. The values can be of any finite data type.

We could express relations of the form "*if the values of some inputs at some given time are X and Y, the value of some output at some other time is Z*". There is a problem with timing in this approach - it is difficult to express at which time points the values are indicated, so it is usually used when the time points are implicitly given (e.g. in synchronous systems) [19].

Another idea is to describe changes of inputs and outputs, because changes occur at single time points. But if we describe the behavior solely in terms of such changes (called events), even descriptions of simple systems can become complicated, as we have to describe all events, even if we are interested only in some of them [14].

We decided to take a mixed approach. We observe the behavior as relations over entities we call system actions. A system action consists of an event (a change of some input or output) which determines the time point when the action is observed, and a set of values of inputs and outputs indicated at the time when the event occurred.

If there are no values indicated, the action is equal to an event. If some event is used as a default (e.g. a change of clock), the actions describe only values. This means that it is possible to use actions for both approaches mentioned above, so we don't lose any expressivity compared to them.

Why did we do all this only to add timing information to observed values, why didn't we simply add timestamps to the values? We wanted the description of the system behavior to be independent from the timing constraints. This means that the function of the system is defined over discrete time points given by the events in system actions, like in synchronous languages [16] or temporal logic [19]. The quantitative timing constraints are given by relations over the set of these events [7, 8] and are independent from the functional description.

We require that the behavior of the system must remain exactly the same after the timing constraints are added. If the timing constraints cannot be satisfied for all sequences of actions allowed by the functional description, we consider the specification to be inconsistent [7]. This means that the timing may not constrain the language defined by the functional description.

The result of this approach is the possibility to verify system timing and function independently. If we know that the function is OK, we only have to check that the timing is consistent (i.e. that the timing constraints do not contradict each other) and that that it satisfies the constraints given in the timing specification [7, 8, 10, 12, 13].

Most models used today do not work like this. When constraints are added to the description, the function is modified, and the constraints do not allow some sequences of events to occur [4, 5, 9, 19]. This often leads to problems with state explosion (e.g. in models based on timed automata, the state space grows exponentially in the number of timers [6]).

Problems with complexity appear again when we try to decide how to describe the relations between inputs and outputs. Every real digital system has finite memory (and thus, finite state space), and every real digital system can be observed only for a finite time and the signals in it are propagated with finite speed over links with finite bandwidth. This means, that there is only a finite number of changes that can be observed in a given time interval, and the system can be described as a finite state machine. As a result, all we need is a language that can describe finite state machines.

But if we use a finite state machine to describe the behavior of a digital system, the description grows extremely large, is not practical, and hard to comprehend to the designer. We also wanted to close parts of the system behavior to a well defined entity, so that they can be reused. For this purpose, we created so-called system agents, which describe the communication of the system with its environment between two well-defined states. System agents can be hierarchically combined by processes based on operations from the agent algebra into more complicated agents. In combination with a starting mechanism that allows asynchronous starting of agents and processes, the behavior of quite large digital systems can be described, while still keeping the description reasonably small.

In the process of designing of the specification, we can start with a general high-level agent, which can be replaced by a process equivalent to a finite state machine, in which every state transition is described by an agent. These smaller agents can be replaced by processes again, until the specification is sufficiently detailed and can be implemented [3].

To summarize, we needed a model that would:
- be a behavioral model not closely tied to events or values
- describe the function of the system independently from timing, but at the same time allow the timing information to be described easily
- allow hierarchical description and reuse of already specified behaviors and i/o communication patterns
- allow to transform the description into a finite state machine easily and efficiently

Although there are models that satisfy one or more of these requirements, we could not find any model that would satisfy all of them. That is why we designed the model described in the next section.

3. THE SPECIFICATION MODEL

3.1 Basic concept of the model

We will look at the system as a black box with a set IV of observable inputs (input variables), set OV of outputs (output variables), and a set SV of some internal state variables (Fig.1). The variables may be of any finite data type. All data type value domains are extended by a symbol u that represents an unspecified value.

```
Inputs {
   BOOL Reset; DATA D;
}
Outputs {
   DATA D; ADDRESS A;
}
State Vars {
   ADDRESS PC;  DATA[32] regs;
   INSTRUCTION IR;  IFORMAT iformat;
}
```

Figure 1. The declaration of inputs, outputs and state variables of DLX in HSSL.

In the system behavior, the situation at some time point will be characterized by an input vector, an output vector, and a state. We denote these vectors and the state as ordered tuples of values of the respective input, output and state variables. The sets of input vectors, output vectors (e.g., see in Fig.2) and states are denoted by the symbols V, H and Q, respectively.

```
1.                          ov: A=PC;
2. iv: D=d_instr;  ov: A=PC;
3. iv: D=data1;    ov: A=addr1;
4.                          ov: A=addr1;
5.                          ov: A=addr1, D=data2;
```

Figure 2. The united input-output vectors (iv-ov) contained in the DLX instruction cycle actions (Fig.4) including only variables with specified relevant (here symbolic) values: d_instr, data1, addr1, data2, and the given value of PC that becomes the value of A.

We represent the time as a finite interval TI. A variable X can be described as a function $X: TI \to D_X$, where D_X is a finite domain of values, and there exists a finite set of rational values $\{a_0,a_1,...,a_n\}$, $a_0 < a_1 < ... < a_n$ from $TI=\langle a_0,a_n \rangle$, and X is constant on $\langle a_i,a_{i+1} \rangle$ for every i=0,...n-1.

3.2   Events and Timing Events

We use events to specify the discrete time points in TI over which the behavior will be defined, and also to express the necessary timing discipline in TI. Events are changes of variable values. We distinguish two basic types of events: up(X,v,i) and down(X,v,i). The event up(X,v,i) is an event where a system variable X changes its value from any value to value v; i is an index that distinguishes changes of X to the same value v in different time points. The event up(X,v,j) is the nearest change of X to v after the event up(X,v,i) if and only if j = i+1. Similarly, down(X,v,i) is an event where X changes its value from v to any other value. We denote the time point at which the event e occurs by tm(e).

To specify discrete time points within the time interval TI we employ so-called timing events. The discrete time is explicitly specified by sequences of timing events which occur during the system behavior (Fig.3). Formally, the discrete time is defined by a function TE → TIR, where TE is the set of timing events and TIR is the set of rational numbers from TI.

```
up(A,PC), up(D), up(A,addr1), up(D,data2)
```

Figure 3. Timing events that specify the discrete time-points in the instruction cycle of DLX.

3.3   I/O words and the Communication set

The behavior of the digital system S in a finite time interval can be characterized by a relation $F \subseteq IQ \times V^* \times H^*$, where IQ is a set of initial states represented by vectors of state variables; V^* and H^* are sets of finite sequences of input and output vectors of the system, respectively. A triplet $(q, r, s) \in F$ is interpreted as follows: If the system is in some initial state q and an sequence $r = v_1 v_2 \ldots v_n$ of input vectors from V is indicated in discrete time points $t_1, t_2, \ldots t_n$, then – as a system response – the sequence s $= h_1 h_2 \ldots h_n$ will occur. We can unite both r and s together with the sequence $e_1, e_2, \ldots e_n$ of timing events that determine the discrete time points $t_1, t_2, \ldots t_n$ into a single sequence $w = (v_1; e_1; h_1) (v_2; e_2; h_2)\ldots (v_n; e_n; h_n)$, where for every i=1,..., n, v_i and h_i in (v_i,e_i,h_i) denote the respective system vectors indicated at the event time $t_i = tm(e_i)$. In every i/o word w the following relation must hold: $tm(e_1) < tm(e_2) < \ldots < tm(e_n)$.

```
action instr_addr {
   te:up(A,PC); ov:A=PC;
}
action instr_in {
   iv:D=d_instr; te:up(D); ov:A=PC;
}
action data_in {
   iv:D=data1; te:up(D); ov:A=addr1;
}
action data_addr {
   te:up(A,addr1); ov:A=addr1;
}
action data_out {
   te:up(D,data2); ov:A=addr1, D=data2;
}
```

Figure 4. Actions of the DLX instruction cycle in HSSL.

We will call a finite sequence of the type w an input/output (i/o) word, and a triplet (v_i, e_i, h_i) an action (Fig.4 and Fig.5). This means, an action is indicated at the time point defined by the timing event with the input and output vector (actions can be compared to actions in CCS [11]).

All possible i/o words create the set of i/o words $W \subseteq (V \times TE \times H)^*$.

```
W_InstrCycle = { StartingAction . instr_addr . instr_in .
FinalAction } ∪ { StartingAction . instr_addr .
instr_in . data_addr . data_in . FinalAction } ∪ {
StartingAction . instr_addr . instr_in . data_addr .
data_out . FinalAction }
```

Figure 5. The set W of the DLX instruction cycle (the action names from *Figure 4* are used instead of vectors).

The relation F can be expressed by a set $CS=\{(q,w); q \in IQ, w \in (V \times TE \times H)^*, (q, proj_V \cdot w, proj_H \cdot w) \in F\}$, where w is an i/o word and $proj_V \cdot w$, $proj_H \cdot w$ denotes the sequences of input and output vectors contained in w. We call CS the communication set, its members (q,w) communications, and the i/o word w from (q, w) communication word.

The semantics of an i/o word w can be described as a couple $A=(Act, R)$, where Act is the set of all its actions and $R = Act \times Act$ is a relation that represents the relative time order of actions. Let $w = (v_{A0}; e_{A0}; h_{A0}) (v_{A1}; e_{A1}; h_{A1}) \dots (v_{An}; e_{An}; h_{An}) = a_0 a_1 a_2 \dots a_n$. If $A = (Act_A, R_A)$ is the semantics of w, then $Act_A = \{a_0 a_1 a_2 \dots a_n\}$ and $R_A = \{(a_0, a_1), (a_1, a_2), (a_2 a_3), \dots, (a_{n-1}, a_n)\}$

3.4 Algebra of sets of i/o words

Based on the semantics of i/o words, we can define an algebra $(2^W, ., \circ, |[S]|, \|, |[ALL]|, w_E)$ where 2^W is set of all subsets of the set W, ".", "$\circ$", "|[S]|","||", "|[ALL]|", are operations of the algebra called sequence, concatenation, parallelism, parallelism with synchronization, parallelism with full synchronization, respectively. The synchronization set S is a set $\{EC_1, EC_2, \dots EC_r\}$. $EC_i \subseteq Act$ for i=1...r. w_E is the empty i/o word whose semantics is $A_{WE} = (\varnothing, \varnothing)$. Definitions of these operations can be found in [0] and are based on [15].

3.5 States and the Final State relation

The behavior of the system can be specified by the relation $F \subseteq IQ \times V^* \times H^*$ or by the corresponding communication set CS. However, we usually work with specifications of some partial behaviors, and the behavior of the whole system is specified in form of a composition of such partial behaviors.

To make such compositions possible, we have to employ states and determine the final state of the system at the end of the execution of a given partial behavior. In other words, besides the relation F, or the set CS, we have to introduce the transition relation fs: $fs \subseteq IQ \times V^* \times Q$

When the system is in an initial state q and the input sequence $r = v_1 v_2 \ldots v_n \in V^*$ is indicated in the time interval TI at discrete time points $t_1, t_2, \ldots t_n$, then a transition to a state q' will be performed, i.e. the state at the time t_n will be q'. We call q' the final state of such partial behavior, and the transition relation fs the final state relation. We denote this by $(q, r, q') \in fs$, and use also the notation $q \xrightarrow{r} q'$.

The relation F can be replaced by the corresponding communication set CS, and the final state relation fs may be expressed in the form $fs' \subseteq IQ \times W \times Q$, where W is the set of i/o words. If $(q, w, q') \in fs'$, then $(q, proj_{V^*} w, q')$ is also in fs, so we will further denote fs' simply by fs. In the case of a deterministic system, these relations become functions. In the HSSL language, the final states are described explicitly (see Fig.6 - FS).

3.6   Agents

To specify a partial behavior of the system in discrete time points and the timing discipline over continuos time, we introduce agents. An agent (e.g., see the DLX agent Fetch in Fig.6) comprises the communication set, the final state function, and timing rules. Formally, the agent is a tuple (X, Y, Z, IQ, TE, CS, fs, TR), where X, Y, Z are sets of input, output and state variables, IQ is the set of initial states, TE is the set of timing events, CS is the communication set, fs is the final function and TR is a set of timing rules.

The communication set CS, which is potentially infinite and describes a set of finite deterministic communications, is described by communication formulae. The syntax of a communication formula is:

$F := s . G \mid G$

$G := a \mid G.H \mid H|H \mid if(c)H$ else $H \mid H\Delta H \mid G|[S]|G \mid G||G \mid G|[ALL]|G \mid G^* \mid G+$

$H := G \mid f \mid a$

where a, $f \in Act$, c is a predicate, F, G are communication formulae, and $S = \{EC_1, EC_2, \ldots EC_l\}$, $EC_i \subseteq Act$, is a synchronization set. Symbols $|, \|, |[S]|$, $\Delta, *, +$, "if (c) − else" represent operators, ALL is a symbol denoting synchronization set S=Act. The actions s and f are called the starting and the final action. Their timing events are denoted by e_f and e_f, respectively.

The semantics of a given communication formula is defined by a labeled transition system $LTS =_{def} (\Sigma, Act, \{ -a\rightarrow; a \in Act, t, s \in \Sigma\})$, where Σ is a set of states, Act is a set of actions and $\{ -a\rightarrow \}$ is a set of triplets (s, a, t), called transitions. The state s is changed by the action a to the state t. The LTS semantics of our specification model closely correspond to that in CCS[11].

The projection of the action sequence and the first state of the transition trace $q_0-a_1\rightarrow q_1-a_2\rightarrow q_2-a\rightarrow...-a_n\rightarrow q_n$ corresponds to a communication word.

Timing rules from the set TR are predicates that describe the timing discipline. To achieve the function specified by the agent A in the defined discrete time points, all timing rules have to be satisfied. Timing rules must not change the function, they can only constrain the occurrence time of actions in time. They specify the relative time difference between two or more events and their description can be found in [0] or [7].

Any type of timing rule can be used in conjunction with a condition c in the form if (c) TR, where TR is a timing rule. The rule TR is applied only if the condition c becomes true.

We say that a particular agent A with a communication set CS is being executed or runs, if some communication word w, $(q, w)\in CS$ is observed and the final state q' is given by the final state function $q' = fs(q,w)$.

```
Agent Fetch {
    CS { StartingAction . instr_addr . instr_in .
        FinalAction }
    FS {  IR = d_instr; PC = PC + 4;
         iformat = switch on (d_instr.opcode) {
                     case 0x0: R_I;
                     case J, JAL: J_I;
                     otherwise I_I; }        }
    TR { aftn(instr_addr, instr_in, TFetch }
    action instr_addr { te:up(A,PC); ov:A-PC; }
    action instr_in { iv:D=d_instr; te:up(D); ov:A=PC;}
}
```

Figure 6. Agent Fetch.

3.7   Composition and Algebra of Agents, Processes

Compositions of agents are defined by means of an algebraic system (AG, ; , |[S]|, |, Δ, if (c) else, loop, c*P, P*c, NIL), where AG is the finite set of agents defined on a system, c is a predicate, S is a set of actions, and the symbols ; , |[S]|, |, Δ, if (c) else, loop, c*P, and P*c denote concatenation, parallelism, branching, disabling, conditional branching and cycle operations of the algebra, respectively. NIL is a null agent, whose execution time is 0, i.e. its communication formula consists just of one i/o vector $(u; e_f; u)$.

Operation disabling A Δ B cancels the execution of agent A if agent B has been started. This operator is used to describe asynchronous interrupt, resets, etc. Loop is equivalent to the recursive definition $B = \mu X . (A; X)$ [17], and c*P, P*c are cycles. Note, that unlike in process algebras and languages based on them [11, 17], we do not employ recursion, because as we stated in section 2, we want to be able to transform the result to finite state machines easily. For the illustration we will describe only concatenation; all the remaining operators can be found in [0].

Let A and B be agents. If C=A; B then the communication set CS_C is

$CS_C = \{ (q_C, w_C); w_C = w_A \circ w_B, w_A \in W_A, w_B \in W_B, q_C \in IQ_C \}$

Final actions f_A of word w_A, and first actions s_B of word w_B must be compatible (i..e., can be merged). We denote it the compatibility relation by $f_A \approx s_B$). The initial states of agent C are $IQ_C = IQ_A$ and the following condition must hold for the set IQ_B: $FS_A = \{ q'; q'=fs_A(q, w); (q, w) \in CS_A) \} \subseteq IQ_B$. The final state function is $fs_C = fs_A \circ fs_B = fs_B(fs_A((q,w); (q,w) \in CS_C)$ and the timing rules of the agent C are $TR_C = TR_A \cup TR_B$

Compositions of agents are expressed by process formulae (Fig.7). Their syntax is PF := A | P;R | PΔR | P|[S]|R | if (c) P else R | loop P | while (c) P | do P while (c) | 0, where A is an agent, P,R are process formulae, c is a predicate, and 0 is an empty formula. P;R represents sequential behavior, PΔR denotes disabling of P when R is started, P|[S]|R means that P,R are executed in parallel with synchronization on actions from the set S, "if (c) P else R" means that if c is true then P will start, otherwise R will, "loop P" expresses iteration of P and "while (c) P", "do P while (c)" express iteration of P with test on condition c at the beginning or end of every iteration.

```
Process Global { Reset; loop { Fetch; Exec }   }
```

Figure 7. Process Global. It describes overall function of the DLX processor. "Exec" denotes an agent that specifies (executes) the behavior during the instruction execution phase.

3.8   The starting mechanism

There are three ways to start an agent:
-     by an event: a function Stf(A) = (STE, PAR) is defined for an agent or process A (Fig.8, "eon" is the event "occurrence of the energy supply"). The starting expression STE is a couple (e, c) where e is an event and c is a predicate. A is started if c is true and the event e occurs. PAR contains agents or processes which may run in parallel with A. When A is started and an B∈PAR is running, A and B will run in parallel. If B∉PAR, then A will be started and B is stopped
-     by a process: the agents execution follows automatically the composition of agents given in the process formula.

```
Starting structure { Stf Global: eon or up(reset,1); }
```

Figure 8. Starting mechanism of DLX; here the set PAR is empty.

3.9 Formal model of the system

A digital system can be described by a tuple S=(DT, IV, OV, SV, V, H, Q, IS, AG, PR, Stf), where DT is the set of data types, IV and OV are sets of input and output variables, SV is the set of state variables, V, H, Q are sets of input vectors, output vectors, and states, AG is a non empty set of agents, PR is a set of processes, and Stf is a starting function.

The semantics of the system specification will be defined as a system process which represents all possible behaviors of the system and is started by the timing event e_{on}. The process finishes after its final action (i.e. if there is no next action by which it could continue) or prematurely by switching off the energy (event e_{off}). Behavior of the system process is delimited by events e_{on} and e_{off}, i.e. to a finite time interval. This property (finite processes) has been already implied by the agent semantics of processes, where agents had been understood as finite time behavior specification entities. This is sufficient for description of practical digital systems.

4. CONCLUSION

We have created a specification model which introduces some non-traditional ways to describe digital systems. We believe that the mixed event-value paradigm results in more concise designs than traditional event based models. The concept of agents and processes as independent partial behaviors allows the designer to reuse parts of older designs in new ones.

As the model is based on finite state machines, verification of its function should not be problematic, especially considering the fact that the timing constraints are clearly separated from the functional description. Verification of timing was already described [8, 7], verification of function is currently being worked on and will be based on checking of equivalence of agents.

We have defined a specification language based on the model, called HSSL. Description of the HSSL language, including example specifications can be found in [0]. A simulator and an editor is currently being developed, together with a graphical user interface for manipulation of communication and process formulae.

We started the development of a compiler from HSSL to a synthesisable subset of VHDL (the first experimental implementation is ready) and we prepare a methodology that would allow to derive a lower bound of power consumption from high level system descriptions.

This work has been supported by the Slovak Scientific Grant Agency under grant No. 1/4290/97 and No.1/7607/20, and is also a part of the European Community INCO Project 977133 VILAB.

REFERENCES

[0] Fristacky N., Kacerik J., Bartos T.: Specification model for description of reactive digital systems and the HSSL specification language. Technical report, available on the web as http://www.dcs.elf.stuba.sk/~nfris/index.html

[1] Harel D. et al: STATEMATE: A working environment for the development of complex reactive systems. IEEE Transactions on Software Engineering, Vol. 16, No. 4, April 1990, pp. 403-413.

[2] Narayan S., Vahid F., Gajski D.: System specification with the SpecCharts Language. IEEE Design & Test of Computers, December 1992, pp. 6-13.

[3] Fristacky N., Kacerik J.: The specification model and language HSL for digital systems. In proceedings of Baltic Electronic Conference'96, Tallinn, 1996.

[4] Alur R., Dill D.: The Theory of Timed Automata, Lecture Notes in Computer Science 600, Springer Verlag, 1991, pp. 28-73.

[5] Alur R., Itai A., Kurshan R., Yannakakis M.: Timing Verification by Successive Approximation, Lecture Notes in Computer Science 663, Springer Verlag, 1993, pp. 137-150

[6] Balarin F., Sangiovanni-Vincentelli A.: A Verification Strategy for Timing-Constrained Systems, Lecture Notes in Computer Science 663, Springer Verlag, 1993, pp. 151-163.

[7] Bartos T., Fristacky N.: Verifying Timing Consistency in Formal Specifications, IEEE Design and Test of Computers, No. 1., Vol. 13, Spring 1996

[8] Cingel V.: A graph based method for timing diagrams representation and verification, Lecture Notes in Computer Science 683, Springer Verlag, 1993, pp. 1-14

[9] Courcoubetis C., Dill D., Chatzaki M., Tzounakis P.: Verification with Real-Time COSPAN, Lecture Notes in Computer Science 663, Springer Verlag, 1993, pp. 274-287

[10] Hulgaard H.: Timing Analysis and Verification of Timed Asynchronous Circuits, PhD Thesis, University of Washington, Seattle, 1996

[11] Milner R.: Communication and Concurrency, Prentice-Hall, 1989

[12] Ti-Yen Yen, Ishii A., Casavant A., Wolf W.: Efficient algorithms for interface timing verification, Proceedings of the EuroDAC'94 conference, pp. 34-39, IEEE Computer Society Press, 1994

[13] Walkup E. A., Borriello G.: Interface Timing Verification with Combined Max and Linear Constraints, Technical Report 94-03-04, University of Washington, Seattle, 1994

[14] Kwang Il Park, Kyu Ho Park: Event Suppression by Optimizing VHDL Programs, IEEE Transactions on CAD of Integrated Circuits and Systems, Vol. 17, No. 8, pp. 682-691, August 1998

[15] Bartos T.: The Semantics of a Mixed Event-Value Specification Model for Digital Systems, Proc. of the ECS'97 conference, Bratislava, September 1997, pp. 101-104.

[16] Berry G.: The Esterel v5 Language Primer, Version 5.10, release 1.0, Ecole des Mines and INRIA, ftp://ftp-sop.inria.fr/meije/esterel/papers/primer.ps.gz

[17] Bolognesi T., Brinksma E.: Introduction to the ISO Specification Language LOTOS. Computer Networks and ISDN Systems, Vol 14, No 1, 1987

[18] Hennessy J. L., Patterson D. A.: Computer Architecture: A Quantitative Approach. 1990. Morgan Kaufmann Publishers, Inc., San Mateo, CA. Second edition, 1995.

[19] Alur R., Henzinger T. A.: Logic and Models of Real Time: A survey, Lecture Notes in Computer Science 600, Springer Verlag, 1992, pp. 28-73.

JESTER

An ESTEREL-*based Reactive* JAVA *Extension for Reactive Embedded Systems Development*

M. Antoniotti†, A. Ferrari†, A. Flescat†‡ and A. Sangiovanni-Vincentelli†∗
PARADES E.E.I.G. Rome, Italy †
D.I.S., Università di Roma "La Sapienza", Rome, Italy ‡
Dept. EECS, University of California at Berkeley, U.S.A. ∗

Abstract. A "programming"language is viewed as a *concrete syntax* for a given semantics. A well defined sematics eases the design of complex embedded systems. JESTER (JAVA-ESTERel), an extension of JAVA based on *synchronous languages*, has been conceived to ease the task of specifying and designing embedded systems. In this paper, the main features of JESTER are described, and discussed in the wide context of reactive embedded system design alongside an example of automotive origin. JESTER is also placed in context with respect to competing approaches and the emerging Embedded and Micro Edition JAVA Specifications.

1. Introduction: Concrete Syntaxes for Precise Semantics

There is an ongoing effort in the system design community to assert the need of very precise *semantics* foundations for any specialized language to be used for embedded devices development [6]. A language is therefore viewed as a *concrete syntax* for a given semantics. As such, it may be quite different on the surface, as long as the underlying semantic model is respected. However, this need is best addressed by taking into account the current software practices and available tools. Therefore we claim that a language for embedded system design should:

− be close to a well-known programming language that is easy to use;

− be expressive enough to incorporate both data and control flow information;

− have a well defined semantics to favor design and verification.

As pointed out in [9], JAVA already has features that make it usable for "embedded system programming". In addition to being a well-known and easy to use language, JAVA has *threads*, *locks* and *synchronization* primitives for concurrency control. However, most of the embedded systems are reactive systems [1] and the underlying semantics and syntactic constructs of JAVA are such that it is difficult to express *reactions* to signals. In addition, JAVA threads are scheduled by the JAVA Virtual Machine (JVM) interpreter so that the actual concurrency control is hidden from the designer. Finally, the overall application size of embedded programs written in JAVA tends to be large, due to the presence of the JVM[2]. To cope with this problem, SUN published the EMBEDDED JAVA Specification (EJS) [11] and the more recent JAVA 2 *Micro*

---

[1] "A *reactive system* (as opposed − as in [3] − to *interactive* ones) which continuously react to stimula coming from the environment by sending back other stimula. Process control systems are representatives of this class.
[2] It must be noted that very small JVMs (in the order of tens of Kbs) *are* available.

205

P.J. Ashenden et al. (eds.), System-on-Chip Methodologies & Design Languages, 205–214.

Edition with the goal to "*address the software needs of dedicated-purpose embedded devices*". The objective is to deliver applications with a very small memory footprint. Nevertheless, we believe that the semantics model and its implications are still not well addressed.

To cope with this problem we decided to develop JESTER (JAVA-ESTERel), an extension of JAVA based on *synchronous languages* and ESTEREL [3] in particular. JESTER fits naturally the requirements of EJS, while providing a strong semantics constructs for the design of embedded systems.

Related Work. There are many other projects aiming at the extension of JAVA with reactive semantics (e.g. [4, 9, 12]). Reactive extensions of C have been proposed as well [8]. Reactive C extensions are usually less radical in design; they seek for an embedding of "reactive primitives" in (possibly legacy) C code. Working with JAVA allows a greater freedom, inasmuch the amount of embedded legacy code is limited.

The aforementioned JAVA extensions are all completely JAVA-based, i.e. they build a layer of classes on top of JAVA the regular JAVA environment. The approaches can be grouped in two broad categories. Either they construct a sort of "simulation" reactive engine on top of JAVA, or they construct a new scheduling machinery based on the Thread interface, which provides some reactive primitives for synchronization.

On the other hand, JESTER takes the road of incorporating the full ESTEREL semantics directly (thus obviating the problems of the semantics underlying JAVA threads). Moreover, a reactive JESTER program is compiled to C (either directly or through POLIS), thus producing tight code[3]. All in all, there is no real "simulation" framework in JESTER per se. Finally, through an interface to POLIS, a JESTER program reuses all the machinery supporting the *Globally Asynchronous, Locally Synchronous* model of computation. While this is possible – in principle – also with the other JAVA extensions we saw, JESTER is the first system to make this integration one of its design goals.

Outline. This paper is organized as follows. First we describe the JAVA extensions implemented in JESTER and their ties to the ESTEREL semantics and to the JAVA operational semantics.discuss how JESTER can fit nicely into both the POLIS framework and EJS. We conclude with a description of an example of automotive origin and with remarks about future work needed to make JESTER more usable and flexible.

2. Extending Java

Our aim is to provide a uniform and "familiar" environment for programming reactive systems, where the data-handling and control-flow aspects are fully integrated in the language (as opposed to either ESTEREL or LUSTRE [5]).

ESTEREL is a very well designed language, which provides the means for the *compact* and precise representation of *synchronous* programs. We transpose it into a JAVA syntax, while maintaining its semantics. The effect is to provide very well behaved *synchronous islands* within

---

[3] Modulo the problems arising from the product of FSMs.

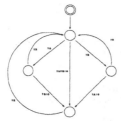

Figure 1. The Esterel ABRO example as a Mealy machine. Standard notation indicates input and output signals.

a JAVA application. Please refer to [3] and to [1] for a complete definition of the programming languages.

2.1. A SIMPLE EXAMPLE

The first example in the ESTEREL primer is the ABRO program.

```
module ABRO:
input A, B, R;
output O;

    loop
        [await A || await B];
        emit O
    each R
end module
```

The program represents the Mealy machine in Figure 1. In order to actually execute the above program, an ESTEREL user must write some C glue code to *embed* the system into a larger one. To execute or simulate the behavior of an ESTEREL program, a user must link against a library which provides the necessary C glue code to tie in the environment or the ESTEREL simulator. The ABRO program in JESTER is depicted in Figure 2. The JESTER processor will produce and link all the necessary JAVA and C glue code for execution and/or simulation.

2.2. JESTER PROGRAM STRUCTURE

JESTER programs may contain reactive extensions or not. If not, they are handed over to the JAVA tool-chain.

In JESTER we make the following choice: the interface of an ESTEREL module is moved at the *class* level, while the behavior of a module is relegated in a special method form (appropriately called **reaction**). Therefore an ESTEREL module corresponds to a JESTER (JAVA) class.

Field declarations which have either the **sensor**, **input**, **output** and **inout** modifier plus the keyword **signal** preceding them are interpreted as *public signal declarations*.

```
import jester.runtime.*;

public class ABRO {
  public input Signal A, B, R;
  public output Signal O
    = new signal() {
          public onEmission() {
            /* Arbitrary Java Code. */
          }
      };

  public void reaction() {
    loop {
      parallel { await(A); await(B); }
      emit(O);
    } each(R);
  }
}
```

Figure 2. The ABRO example in JESTER.

Signals have a double representation in JESTER. One as a first-class data type in the extension and the second one as standard JAVA classes used by the runtime environment. Specialized *valued signals* can be obtained by extending these classes.

The runtime output signal classes contain a method called **onEmission** (*cf.* Figure 2). This method is used to attach arbitrary code to emissions in a much more direct and seamless way than in the case of ESTEREL, where the interface must be specified as a separate C definition.

Method reaction. Each "reactive" class *must* implement a method called **reaction** with signature

```
public void reaction ();
```

Such a method is used as a *syntactic anchor* for the set of tools provided with JESTER. This method is the only one which will be allowed to contain *reactive* statements.

JAVA *Arrays.* ESTEREL lacks direct support for arrays. JESTER relies on JAVA syntax to define them. E.g.

```
double[] matrix[];
```

is recognized by the JESTER translator, and appropriate type declarations and specialized function calls are thus generated.

```
import jester.runtime.*;

public class prodcon {
  producer P = new producer();
  consumer C = new consumer();

  public void reaction () {
     signal void S;

     parallel {
       P.reaction(P.O = S);
       C.reaction(C.I = S);
     }
  }
}
```

Figure 3. The JESTER producer and consumer example.

2.3. JESTER EXPRESSIONS AND STATEMENTS

All ESTEREL constructs are translated into JESTER (with some limitations). Please refer to [1] for a complete discussion of the syntax and semantics. Figures 2 and 3 should give the reader an idea of JESTER actual syntax. In the following we discuss only the special cases of the *renaming operation*.

2.3.1. *The run Statement*

Calls to the reaction method are recognized by the JESTER preprocessor and are translated into proper ESTEREL run statement. Local signals become the *wires* connecting different sub-modules.

E.g. consider the producer/consumer example in JESTER presented in Figure 3 (producer and consumer are straightforward: they just emit an output and await on an input).

The actual argument list of the *reaction* method is treated specially by the JESTER translator. Assignments to signals within the reaction method argument list, are recognized and transformed in the proper ESTEREL "renaming" operations.

3. Tools and Implementation

The JESTER tool-set produces a set of intermediate forms amenable to automatized synthesis of embedded systems. Figure 4 shows the overall architecture of the JESTER processor. Note that adding a new module (e.g. one producing code for ocjava, *cf.* www.esterel.org) is a relatively simple matter.

210

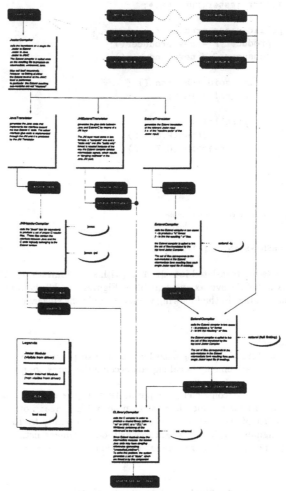

Figure 4. The current architecture of the JESTER translator. All the submodules are depicted. The dark rectangular ones are modules which are available to the "top level" driver, be it a *command line* or a *window-based* interface. The diagram also shows the intermediate files produced. This figure is provided as a map for the developer and the reader interested in looking at the code.

A Hardware/Software Synthesis Work-Flow The JESTER tool-set translates a JESTER specification into a set of JAVA, ESTEREL, and C files (mostly containing JAVA/JNI glue code). The production of *Communicating Finite State Machines* (CFSM) [2] is supported by producing SHIFT files, which can be directly input to tools like POLIS.

The choice of the intermediate formats produced by the JESTER translator is dictated by the desire to be compatible with other tools and standards.

The JESTER translator is implemented in JAVA plus some helper scripts. Therefore it runs on most platforms. Moreover, the `jester` translator has the same "use flavor" of `javac`, i.e. all dependencies are automatically taken care of, by forcing the compilation of source-only modules

The Personal/Embedded JAVA/JESTER Work-Flow The results of the JESTER compiler can be used to produce a (embedded) JAVA application which can be linked with the results of the compilation of the ESTEREL intermediate form. We have two goals: (a) to reuse already existing tools as much as possible and (b) to produce Java code which can be downloaded on various HW/SW platforms like those which will support the EMBEDDED JAVA specification [11].

Beside the reuse of POLIS, we are currently monitoring various industry efforts to produce a native JAVA compiler[4] to fill in the last steps of the EMBEDDED JAVA work-flow, and various *real-time operating system*, third parties JVMs and RTOSes to produce a proof of concept complete path-to-implementation for *Personal and Embedded JAVA Specifications* (PJS and EJS) applications, with the inclusion of specialized classes representing *hardware* devices.

The JAVA code produced by the JESTER translator passes the `JavaCheck` test for the PJS 1.1 and it can be compiled with the PJS compiler version 3.0 (*cf.* the JAVA web pages for this tool). Of course this is only significant if we can ensure that the C code produced by the ESTEREL compilers satisfies the constraints imposed by the target HW platform. More experimentation is needed on this specific aspect before making and supporting more precise claims. However, we can safely state that the current JESTER runtime library (implemented in JAVA) complies with the PJS/EJS restrictions.

Given the above considerations, we believe that there is an opportunity to make a significant contribution to the PJS and EJS effort, *since our approach frames a PJS or EJS JAVA application within a clear semantics model.*

3.1. OPEN PROBLEMS

There are some aspects of JAVA programming that still need more in-depth considerations: garbage collection and threading. A third aspect that is still open for investigation is the *communication refinement* problem.

Garbage Collection. Although GC technology has improved [7] over the years, we believe that some investigation into the use of automated static memory allocation would be worthwhile.

Threading. The JAVA Threads model does not directly fit within a reactive framework. The work by Passerone et al. [9] (the JAVA PureSR package), builds upon the JAVA threads package, while aiming at substituting it. A similar approach is taken by Young et al. in [12].

---

[4] The JAVA to gcc tool suite translator from Cygnus is a candidate – http://www.cygnus.com.

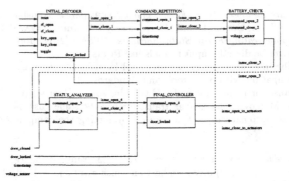

Figure 5. The block diagram for the door lock example. Each block is a separate JESTER class (or module) and the system is composed synchronously according to ESTEREL semantics.

Communication Refinements. This problem is related to the threading one. Right now, a user has little or no support from the tool for checking semantics constraints when combining reactive JESTER classes. Only using POLIS the user has some support for *globally asynchronous, locally synchronous* (GALS) systems, but the current interface is not stable. This is a rather complex theoretical problem, which is related to the problem of *decomposability* of specifications. We have no defined solution for it yet.

4. A More Complex Example

We used JESTER to implement a specification of an automotive centralized door lock system obtained from Magneti-Marelli S.p.A, Italy [10]. The original specification was given in natural language plus some timing diagrams. Its aim is to handle a key and infrared inputs in order to activate the locking and unlocking of a car doors. The overall system is depicted in Figure 5. Figure 6 shows a snippet of the `InitialDecoder` module (a simple one for lack of space).

At its core, the specification required that each *lock* or *unlock* command issued by the user should be executed only if certain conditions were met. E.g the battery voltage must be above a 9 V threshold, the frequency of the commands could not be too high (a "debouncing" module was thus introduced), the final actuator command is to carried out after a given delay, etc.

We "simulated" the JESTER code by embedding it into a simple JAVA driver whose main aim was to link in a data collection and experiment management JAVA library we developed. The data collected can be manipulated by, e.g. a regular spreadsheet. A sample output of one of the experiments we ran is shown in Figure 7.

```
import jester.runtime.*;

public class InitialDecoder implements Reactive {
    public input signal void rfOpen;
    public input signal void rfClose;
    // ... other input and outputs go here.
    public output signal void issueOpen;
    public output signal void issueClose;

    public void reaction() {
        parallel {
            loop {
                await(rfOpen || rfClose);
                present(rfOpen) { emit(issueOpen); }
                present(rfClose) { emit(issueClose); }
            }
            // ... other loops on other input signals go here.
        }
    }        // end of reaction method
}            // end of InitialDecoder class
```

Figure 6. A snippet of the InitialDecoder module, which is in charge of translating the user inputs to an internal representation for the door lock system. A number of reactive loops are made to wait in parallel for user inputs. Once one of the loops detects an input it issues an "internal" command represented by either a issueOpen or issueClose signal.

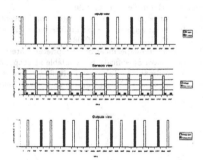

Figure 7. A sample of the data produced by the JESTER coded door lock system. The graphs have been produced via a well known spreadsheet program. The first graph represents a sequence of inputs of the infrared toggle command. The second graph shows the reading of the voltage sensor plus the reading of the "door locked" sensor. Finally the third graph shows the issuing of the actuator commands (an alternation of "open" and "close" commands at the correct and delayed times.

5. Conclusion

We described the JESTER system as a set of extensions to JAVA, making this language more useful for the design of reactive and embedded systems. These extensions are in the spirit of the ESTEREL language and are supported by a set of tools. In particular, we specify a *synthesis work flow* that relies on tools like the ESTEREL system, and like POLIS in order to produce optimized code for embedded systems.

Finally, since our approach is well integrated with the current EMBEDDED JAVA specification, it will enjoy a wider acceptance by the Java community.

References

1. M. Antoniotti and A. Ferrari. Jester project home page. Available at http://www.parades.rm.cnr.it, in the "Software and Projects" section, PARADES E.E.I.G. Rome, ITALY, 1999.
2. F. Balarin, M. Chiodo, P. Giusto, H. Hsieh, A. Jurecska, L. Lavagno, C. Passerone, A. Sangiovanni-Vincentelli, E. Sentovich, K. Suzuki, and B. Tabbara. *Hardware-Software Co-design of Embedded Systems – The POLIS Approach.* Kluwer Academic Publishers, 1997.
3. G. Berry. The ESTEREL v5 Language Primer. Available at http://www.inria.fr/meije/esterel, I.N.R.I.A. Sophia-Antipolis, FRANCE, 1997.
4. F. Boussinot, G. Doumenec, and J.-B. Stefani. Reactive Objects. *Annales des Telecommunications,* 51(9–10):459–473, 1996.
5. P. Caspi, D. Pilaud, N. Halbwachs, and J. Plaice. LUSTRE: a Declarative Language for Real-Time Programming. In *Proc. Conf. on Principles of Programming Languages,* 1987.
6. T. Henzinger, L. Lavagno, E. Lee and A. Sangiovanni-Vincentelli, and K. Vissers. System Level Design Languages: Abstract Syntax, Syntax Transformation and Semantics. Private communication, April 2000.
7. R. Jones and R. Lins. *Garbage Collection: Algorithms for Automatic Dynamic Memory Management.* John Wiley & Sons, 1996.
8. L. Lavagno and E. Sentovich. ECL: a Reactive Extension of C. presentation in the *Forum on Design Languages* 1998, Lausanne, Switzerland, September 1998.
9. C. Passerone, R. Passerone, C. Sansoé, J. Martin, A. Sangiovanni-Vincentelli, and R. McGeer. Modeling Reactive Systems in Java. In, *Proceedings of the Sixth International Workshop on Hardware/Software Codesign,* pages 15–19, Seattle, WA, U.S.A., March 1998.
10. Magneti Marelli S.p.A. Door lock system specification. Proprietary internal document, 1998.
11. Sun Microsystem, Inc. Embedded Java Specification. Available at http://java.sun.com, 1998.
12. J. S. Young, J. MacDonald, M. Shilman, A. Tabbara, P. Hilfinger, and A. Richard Newton. Design and Specification of Embedded Systems in Java Using Successive, Formal Refinement. In *Design Automation Conference,* 1998.

A four-phase handshaking asynchronous controller specification style and its idle-phase optimization

Rafael Kazumiti Morizawa and Takashi Nanya
Research Center for Advanced Science and Technology, The University of Tokyo

Key words: Asynchronous circuit, asynchronous specification, logic synthesis, CAD tool

Abstract: A known problem of the four-phase handshaking protocol is that the signals
               involved in the handshake need to return to its initial state (a return-to-zero
               phase, also known as idle-phase) before starting another cycle, in which no
               useful work is usually done. In this paper we first define an easy-to-write
               specification style to specify four-phase handshaking asynchronous
               controllers that can be translated to an STG to obtain a gate-level
               implementation using existing synthesis methods. Then, we propose an
               algorithm that takes a controller specification written using our specification
               style and finds an optimized timing to start the idle-phase such that its gate
               level implementation has the idle-phase overhead reduced.

1.   INTRODUCTION

In the last few years we have seen the development of various methods
and tools to synthesize asynchronous controllers. There are basically two
different approaches; one uses high-level specification languages, whose
main characteristic is the powerful expression of concurrency and are based
on CSP [Berkel93c,Martin87]; and the other uses specifications such as
signal transition graphs (STG) [Chu87] or finite-state machines (FSM)
[Nowick93a,Yun94].

Specifying a complex controller behavior using high-level specification
languages is not very difficult. However, specifying such behavior in the
form of STGs or FSMs is a very hard and error prone work. Thus, it is

215

P.J. Ashenden et al. (eds.), System-on-Chip Methodologies & Design Languages, 215–229.

necessary to make a bridge between a high-level specification language and lower level abstraction formats such as STGs of FSMs. There has been extensive work trying to bridge this gap both in the synchronous [Helbig95] and asynchronous [Kudva95] domains. None of these approaches, however, generates an STG from a high-level specification. The work described in [Blunno98] defined a subset of the Verilog-HDL such that a designer could write a specification that can be simulated in a Verilog simulator and converted to an STG. Inspired by this work, we have defined a Verilog HDL subset and a specification style to specify four-phase handshaking protocol controllers.

A known problem of the four-phase handshaking protocol is that a return-to-zero phase (also called idle-phase) of the signals involved in the handshake is necessary before starting a next cycle, during which no useful work is usually done. In [Furber96,Martin97], the synthesis of controllers trying to minimize the overhead of the idle-phase were proposed. However, in their work there was no proposal of an automated procedure to minimize the overhead of the idle-phase of general controllers. In [Kondratyev99a], a methodology were proposed for optimizing area, performance or power of controllers specified using STGs. That methodology accomplishes it by selecting the order in which some of the idle-phase events occur.

The idle-phase overhead in four-phase handshaking protocol controllers can be reduced if we obtain an implementation of the controller that starts the idle-phase as soon as the working phase has finished, with maximum concurrency with the other signal changes. We show that an implementation obtained from such specifications may not reduce the idle-phase overhead.

In this paper we propose an easy-to-write specification style for four-phase handshaking controllers, and an algorithm that finds an optimized timing to start the idle-phase that reduces the idle-phase overhead. The paper is organized as follows. We first present the specification style, and the algorithm used to translate it to an STG. Then, we show the approach we use to systematically decrease the overhead of the idle-phase. Next we show an experiment comparing our algorithm with the methodology presented in [Kondratyev99a]. Finally we present the conclusions of this paper.

2.   CONTROLLER SPECIFICATION

We have defined a subset of the Verilog HDL that can specify a four-phase handshaking controller with two objectives in mind. The first objective is that the designer must be able to describe the behavior of a controller by specifying which output signals change for each change in one or more input signals; and should be able to simulate the specification (even integrating it within a larger system). The second objective is to be able to synthesize a controller that implements the specification.

We have chosen the Verilog HDL to implement the ideas we propose in this section because it is a specification language that is simple enough to write a controller specification, and to add extensions to it. This does not reduce the generality of the ideas proposed in the paper for we believe it can also be applied to other hardware description languages.

2.1 Terminology

We call a pair of request and acknowledge signals a *handshake pair*. A request or acknowledge signal within the same handshake pair is a *handshake signal*. Given a controller with two or more handshake pairs, a *passive handshake* is the handshake pair that is responsible for starting the controller's operation. An *active handshake* is the handshake pair that activates data path components or other controllers. Suppose a request signal r and an acknowledge signal a constitute a handshake pair. Let us assume that their initial values are zero. In the four-phase handshaking protocol, these signals switch in the following order: $r+ \rightarrow a+ \rightarrow r- \rightarrow a- \rightarrow r+$, where a "+" means a 0 to 1 signal transition and "−" means a 1 to 0 signal transition. We call this signal switching sequence a handshake loop.

2.2 Controller model

The controller model we assume in our controller specification is shown in Fig. 1. In this model we assume that there is one passive handshake pair, and one or more active handshake pairs. For each handshake pair, a handshake signal can be either a 1-out-of-n signal or a 1-bit signal.

2.3 Specification style and grammatical restrictions

2.3.1 Specification style

The designer specifies the behavior of the controller by defining the output signals that change when a set of its input signals change, and by declaring which signals form a handshake pair. The declaration of the handshake pairs by the designer is the key feature of our specification style since it allows the designer to specify only the causal relationships between a controller's input and output signals without adding explicitly in the specification the causal relationships between the signals of a handshake pair.

In [Blunno98], a subset of the Verilog HDL to specify a controller and translation rules to translate the specification to an STG were also defined. The differences between their specification style and ours are the declaration of the handshake pairs and the larger subset of the Verilog HDL that we support. The larger subset of the Verilog HDL allows the designer more flexibility when specifying a controller.

The explicit declaration of the signals that form a handshake pair helps simplifying the way a designer writes the specification since it is not necessary to add in the specification statements that represent the causal relationships between signals within the handshake pair.

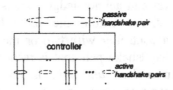

Figure 1 Controller model

2.3.2   Verilog HDL grammatical restrictions

We have defined a subset of the Verilog HDL that allows designers to write a wide range of controller specifications while simplifying the task of translating it to an STG.

The controller can be specified using one or more `always` module items which specify concurrent threads. We have defined that `always` module items must have edge triggered conditions called *event expressions* that set the conditions under which the thread can start its execution. There are two restrictions on how these concurrent threads have to be specified: the first is that the threads are non-reentrant, i.e. a thread has to finish its execution before it is enabled to start another one; and the second is that the signals used within the condition that enables the execution of a concurrent thread must be reset before it is evaluated again.

The initial conditions of the controller are specified using an `initial` module item, which is executed only once within a simulation run.

The statements allowed within a module item are assignments, conditional branches, `wait` statements, sequential blocks and parallel blocks. Assignments are used to make the controller's output signals to rise or fall, and the only values allowed to be assigned are "0" or "1". Conditional branches are specified using `case` statements. A `case` statement must be preceded by a `wait` statement or be the first statement executed in a module item. The `wait` statement halts the execution of a thread until its argument expression is evaluated true. A sequential block is a set of statements nested within begin/end that are serially executed. A parallel block is a set of statements nested within a fork/join that are executed in parallel. A statement following a parallel block can only start execution after all the statements within the parallel block have finished their execution.

The expressions allowed in the argument of a wait statement execute bit operations. The allowed operators are INVERSION, AND, and OR. We have put a restriction on the operators of an OR operator: all the arguments of an OR operator must be mutually exclusive. The argument of event expressions is restricted to unary expressions. There must be an edge identifier, such as posedge or negedge, that indicates which edge, rising or falling, of the unary expression must be observed by the event expression.

2.3.3 Added constructions

The added constructions are specified as comments in the Verilog HDL source of a controller, and declares which signals form a handshake pair.

Handshake pairs are declared using the construction handshake_pair. This construction has two arguments, separated by a comma. The first argument is a list of request signals, and the second argument is a list of acknowledge signals. We have also introduced a variation of the above declaration: handshake_pair*. Handshake pairs declared using this declaration may have the start timing of its idle-phase modified by the algorithm to reduce the idle-phase overhead, proposed in the next section.

An example of a specification is shown in Fig. 2. The piece of Verilog code models the behavior of a sequencer. Signals R and A constitute a passive handshake pair, and signals r and a constitute an active handshake pair.

```
module sequencer(R, A, r, a);
    input R, a;
    output A, r;
    reg A,r;
// handshake_pair R, A;
// handshake_pair r, a;
always @(posedge R)
begin
    r = 1;
    wait(a);
    fork
        r = 0; A = 1;
    join
end
always @(negedge R)
begin
    wait(~a);
    A = 0;
```

```
            end
         endmodule
```
Figure 2. An example specification of a sequencer.

2.4 Translation to STG

The translation from the proposed specification style to an STG is basically a two-step algorithm. In the first step the algorithm determines the usage (reference or assignment) frequency of each variable in the specification. This information is needed to differentiate the various usage instances of a variable, which must be reflected in the translated STG. The second step is the core of the translation algorithm and relies on two basic points:
- The behavior of a controller is described by specifying output signal changes in response to input signal changes.
- The specification includes the knowledge of which signals constitute a handshake pair and the switching sequence of signals in its handshake loop.

The first point allows adding causal relationships from input to output signal transitions. The causal relationships alone are not enough to translate the specification to an STG. The second point provides the remaining information to translate the controller specification to an STG. The knowledge of which signals constitute a handshake pair allows the addition of causal relationships from output to input signals.

TRANSLATION ALGORITHM FROM VERILOG HDL TO STG
Given a Verilog HDL source code do:
Step 1: Determine the usage frequency of variables in the specification
Step 2:
foreach always statement **do**
Build from the event expression an STG G_{event}
Build from the body of the always statement an STG G_{always}
Connect with arcs the STG G_{event} to the STG G_{always}
od
Add the causal relationships of the handshake pairs
Determine the initial marking

In step 1 the usage frequency of variables is necessary because we need to differentiate between the different uses of a same variable that appear in different instances in the specification of a controller. We define below the translation rules used in step 2 of the algorithm described above.

Assignments: A "0" assignment is translated to a falling transition while a "1" assignment is translated to a rising transition.

wait statements: A Verilog `wait` statement is used to synchronize the execution of statements by evaluating its argument expression. The evaluation of its argument is level sensitive. Thus, the statement "`wait(a) b = 1;`" cannot be translated to an STG like "$a+ \rightarrow b+$" since it does not express the behavior of the **wait** statement when the variable a is already high. The translation rules of `wait` statements where the arguments are an unary expression, or ANDed expressions are shown in Fig. 3(a). The translation rule of `wait` statements where the arguments are ORed expressions is shown in Fig. 3(b). The latter translation rule is based on the restriction that the arguments of ORed expressions are mutually exclusive.

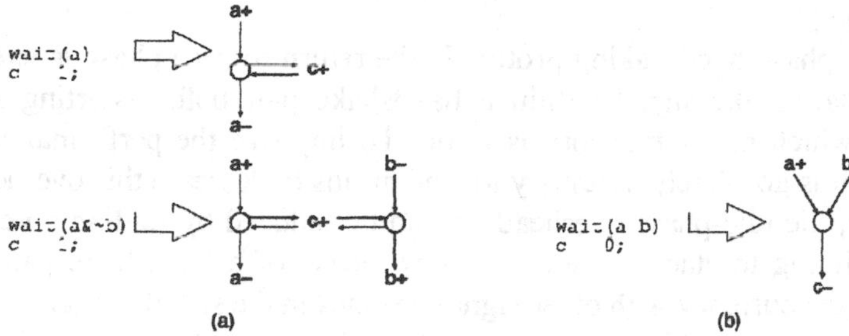

Figure 3. wait statement translation rule when the arguments are an unary or ANDed expression (a); and the translation rule when the argument is an ORed expression.

event expressions: An event expression is used to synchronize the execution of statements by evaluating its argument expression. It differs from a `wait` synchronization because the evaluation of an event expression argument is edge sensitive. Figure 4 shows the translation rules for event expressions.

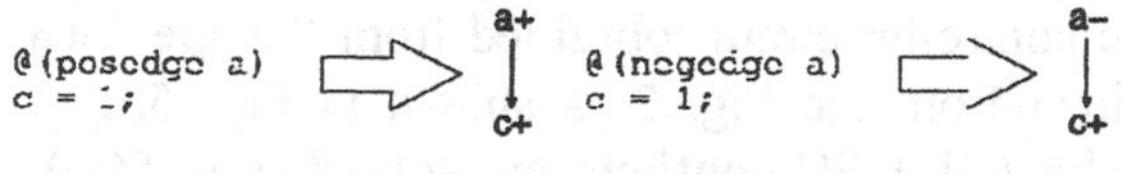

Figure 4. Event expression translation rule.

case statements: `case` statements are translated to an STG by first translating the statements of each conditional branch, and then introducing a place that has arcs going from it towards each of the branch's STGs.

sequential block: After translating each statement of a sequential block, connect the obtained STGs with arcs to form a sequence. The arcs must link input transitions to output transitions, or two consecutive output transitions.

parallel block: After translating each statement of a parallel block, add one dummy event representing a forking event and one dummy event representing a join event (dummy events are going to be eliminated later). Connect with arcs the forking event to the transitions of each translated statement, and these statements to the join event.

The initial marking of an STG represents the initial state of the controller, and are graphically shown as token placed on an STG arc or place (see figures 5(a), 5(b), and 7(a)). The initial marking is determined by using the information about the initial state of the controller (specified in the initial module item) and the handshake pair declarations.

3.   OPTIMIZING THE RETURN-TO-ZERO PHASE

In four-phase handshaking protocols the return-to-zero phase is executed for each handshake signal within a handshake pair before starting a next cycle, in which no useful work is done. To improve the performance of a controller it is absolutely necessary to find means to decrease this overhead.

Ideally, the idle-phase overhead should be reduced by finding an earliest possible timing to start the return-to-zero phase of a handshake pair, with maximum concurrency with other signal changes in the specification.

However, by analyzing typical controller specifications using the proposed specification style, we noticed that starting the idle-phase at its earliest possible timing (with maximum concurrency with the other signal changes) is very likely to introduce a complete state coding (CSC) conflict in the translated STG. An STG has a CSC conflict when in a state graph generated from the STG there are two or more states with the same encoding. An effective way to solve CSC conflicts is by adding internal signals that are used to differentiate the states that have the same encoding. These added signals may, in turn, increase the overhead we want to decrease.

Consider the controller circuit obtained from the specification in Fig. 2. The STG obtained from the Fig. 2 is shown in Fig. 5(a). Using the tool petrify [Cortadella96b], CSC conflicts are solved (Fig. 5(b)). The obtained gate-level implementation is shown in Fig. 5(c). (We have used the complex gates available in   NEC's 0.25 μm standard gate library.) In the translated specification, shown in Fig. 5(a), the transition r−, responsible for the idle-phase start of the handshake pair r/a, is at its earliest possible timing and has maximum concurrency with the other signals.

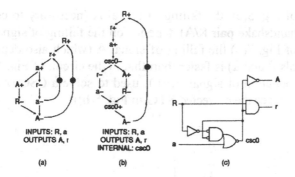

Figure 5. STG obtained from the translation of the specification in Fig. 2 (a); after CSC conflicts are solved (b); gate level implementation (c).

Figure 6 shows a modified version of the specification of Fig. 2. The difference between these two specifications is that the former executes statements "A = 1;" and "r = 0;" serially, while in the latter they are executed in parallel. In the STG of Fig. 7(a), obtained from the translation of the specification of Fig. 6, the transition r− that starts the idle-phase of handshake pair r/a is not maximally concurrent with the other signals changes, since it is not concurrent with the transition A+. We call this fact a *reduction in concurrency*. The gate level controller synthesized from the STG is shown in Fig. 7(b).

```
module sequencer(R, A, r, a);
    input R, a;
output A, r;
    reg A, r;
    // handshake_pair R, A;
    // handshake_pair r, a;
    always @(posedge R)
    begin
        r = 1;
        wait(a);
        A = 1;
        r = 0;
    end
    always @(negedge R)
    begin
        wait(~a);
        A = 0;
    end
endmodule
```
Figure 6. A modified version of the specification of the sequencer.

Comparing the circuit of Fig. 7(b) with the circuit shown in Fig. 5(c), we notice that in both controllers the idle-phase of the handshake pair r/a (the falling of signal r) has roughly the same delay after signal a rises. However, in the circuit of Fig. 5(c), the falling of signal **A** (necessary to complete the idle-phase of handshake pair **R/A**) depends on the falling of signals **R** and **a**. In the circuit of Fig. 7(b) the falling of signal **A** (which also depends on the falling of signals **R** and **a**) is faster than that in the circuit of Fig. 5(c). This is due to the introduction of signal `csc0`, used to solve a CSC conflict in the STG of Fig. 5(a) (see the corrected STG in Fig. 5(b)).

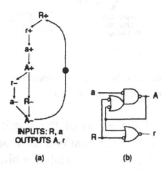

Figure 7. STG obtained from the specification of Fig. 9 (a); gate level implementation (b)..

The reason why there is no CSC conflict in the STG of Fig. 7(a) is because in the specification from which it was obtained the statements "r = 0;" and "A = 1;" are not executed in parallel after the statement "wait(a);" like in the specification of Fig. 2. The serialization of the statements "r = 0;" and "A = 1;" eliminates some states in the resultant STG, including the state that causes the CSC conflict.

Thus, we have a dilemma: if the idle-phase of a handshake pair starts at its earliest possible timing with highest concurrency we may decrease the performance of the obtained controller. One way to solve this problem is to find means to solve some CSC conflicts without adding new signals. This can be done by decreasing the concurrency of the specification. Concurrency can be reduced by modifying the idle-phase start timing of the handshake pairs. However, decreasing the concurrency without any criterion does not help obtaining better performance controllers.

In the criterion proposed in this paper, instead of starting the idle-phase as soon as the corresponding working phase has finished, its start is delayed by one signal transition, i.e., the idle-phase starts **after** an output signal transition. This reduces "slightly" the concurrency of the STG obtained from the specification. We have noticed that this reduction in concurrency is enough to eliminate various CSC conflicts in typical controller specifications.

The algorithm to find the start timing of the idle-phase assumes that the delay of the controller's output signals is lower than the delay of the data path's response. It takes as input a controller specification in which the designer specifies which handshake pairs must have the idle-phase optimized.

For each selected handshake pair, the algorithm first searches for the statement that causes the idle-phase to start. Then, it determines an execution ordering of the statement that starts the idle-phase and the other statements in the specification, following the criterion presented above.

IDLE-PHASE OPTIMIZATION ALGORITHM

Given a Verilog HDL source code do:

foreach handshake pair marked to have its idle-phase start timing modified **do**

Find the statement(s) $S_{ipstart}$ that starts the idle-phase

foreach $S_{ipstart}$ **do**

Remove the $S_{ipstart}$ from the source code

od

Find the statement(s) $S_{ipcause}$ that causes the idle-phase to start

foreach $S_{ipcause}$ **do**

Find the assignment statement S_0 that is executed after $S_{ipcause}$

Find the statement S_1 that is executed after S_0 (and that it is in the same nest level as S_0)

if S_1 exists **then**

Insert $S_{ipstart}$ so that it is to be executed concurrently with S_1

else

Insert $S_{ipstart}$ so that is to be executed immediately after S_0

fi

od

od

The concurrency of the specification of Fig.2 is reduced by serializing the execution of statements "A = 1;" and "r = 0;" which are executed after statement "wait(a);". Consider the handshake pair r/a. The statement that corresponds to the idle-phase start is "r = 0;" and the statement that causes the idle-phase start is "wait(a);", from the relationships of a handshake loop. After removing "r = 0;" from the source code, the assignment statement that is executed after "wait(a);" is "A = 1;". Then, the statement that corresponds to the idle-phase start is placed in the source code so that it is executed immediately **after** statement "A = 1;" to finally obtain the source code shown in Fig. 9.

4.   EXPERIMENT

In this section we show the effectiveness of our algorithm in finding an optimized timing for an idle-phase to start. We test our proposed method in four types of controller specifications described below.

1. A controller that takes the stream of a pipeline and branches it into two other pipeline streams.
2. A controller that merges the stream of two pipelines into one pipeline stream.
3. A benchmark controller developed by our research group.
4. A 2-way sequencer using concurrent protocol [Plana98].

For each controller specification we compare, using simulations, three implementations of the gate level circuits. The implementation obtained using petrify and no concurrency reduction is designated **noredc**. The implementation obtained by using petrify's concurrency reduction methodology is designated **petrifyredc**. The implementation obtained using the proposed concurrency reduction methodology and then synthesizing using petrify is designated **ourredc**.

In the simulation, the first two controllers' are actually controlling the branch and merge of pipeline streams. The other two controllers, a data path that acknowledges each signal from the controller in 1 ns was set up. Table 1 shows the cycle time of the controller's different gate level implementations. The cycle time is the time taken by the controller to execute the handshake loop of the passive handshake which starts the controller operation.

Table 1 The average cycle time in [ns] of the controllers.

controller	noredc	petrifyredc	ourredc
1	6.57	6.65	6.29
2	6.69	4.48	5.44
3	30.55	34.58	28.33
4	5.77	5.00	4.77

Although the number of tested circuits is small, we can see that a specification with high concurrency does not necessarily results in a gate-level implementation with faster cycle time. Despite the reduction in concurrency, the fewer internal signals added to solve the CSC are responsible for the better performance (see table 2). Also, because in our approach the concurrency of the specification is not drastically reduced, since the start of the idle-phase of a handshake pair is moved to a statement that is executed just after it, the impact in the performance due to concurrency reduction is small.

Now our approach (see table 1, column **ourredc**) to optimize the idle-phase is compared with petrify's concurrency reduction approach (see table 1, column **petrifyredc**), which is the implementation of the methodology in [Kondratyev99a]. Although the cycle time of controller 2 is better using petrify's concurrency reduction approach, the cycle time of all other controllers was smaller using our proposed algorithm.

Table 2 The number of signals introduced in the STG to solve CSC conflicts.

controller	noredc	petrifyredc	Ourredc
1	1	0	0
2	2	0	0
3	4	1	1
4	2	0	0

We have this difference in performance because the default behavior of petrify when reducing the concurrency of a controller specification ends up reducing the concurrency of many signals unnecessarily. It may be possible to obtain an STG (and therefore a gate level implementation) with similar concurrency of the ones obtained by our approach by fine-tuning the tool's options (as described in petrify's documentation [Cortadell98]). However, specifying precisely the fine-tuning options to optimize the idle-phase start is a very hard work for designers (especially in complex specifications).

Further experiments with other controller specifications showed that our approach in reducing the idle-phase overhead is more effective than petrify's approach when the controller controls data paths with long idle-phase times, or when the data-paths under control are reused many times within the cycle time of the controller. For controllers that control data paths with very short idle-phase times, petrify's approach showed to be very effective. The reason is that petrify aggressively reduces the concurrency of the controller specification, which results in much simpler and lower latency controller implementations than those obtained using our approach. If the idle-phase time of the data path is very short the impact of the return-to-zero phase in the performance is very low compared to the impact of the latency of the controller implementation.

This experiment illustrates how our proposed idle-phase optimization algorithm finds a solution for reducing the idle-phase overhead of four-phase handshaking asynchronous controllers (in which the delay of its output signals is lower than the delay of the data path's response), without lowering very much the concurrency of the controller specification. The methodology described in [Kondratyev99a] finds a wide range of solutions that may satisfy a given optimization constraint. However, in the particular case of optimizing the idle-phase overhead, setting the optimization parameters in order to fulfill these constraints, while not decreasing very much the concurrency of the STG, may be difficult, especially in complex specifications.

5.   CONCLUSIONS

In this paper a four-phase handshaking controller specification style that allows designers to simulate and, after translating to a STG (using the translations rules presented in this paper), synthesize controllers as been defined. The proposed specification differs from the specification style of [Blunno98] in the point that it supports a larger subset of the Verilog HDL, thus giving the designer more flexibility to specify controllers.

The specification style has been implemented by defining a subset of the Verilog HDL and adding extensions to it. The added extension is also a directive to reduce the idle-phase overhead of selected handshake pairs. We have proposed an algorithm that finds a timing to start the idle-phase of the selected handshake pairs such that an STG obtained from the translation of a specification is not the most concurrent STG that can be obtained. However, its gate-level implementation may be faster than the implementation obtained from a highly concurrent STG due to the reduced number of internal signal that are added to solve CSC conflicts.

There are a few aspects in this work that need further study. Among them are the extension of the controller model supported by this specification style, and an algorithm to choose automatically, without designer intervention, the handshake pairs that should have the idle-phase start timing optimized. The only information that can be extracted from our controller specification style is how the controller's signals interface with the environment, which is not enough to select handshake pairs that should be optimized. Information of the structure of the data path is also needed.

ACKNOWLEDGEMENTS

This work was partly supported by the Semiconductor Technology Academic Research Center (STARC) and by the Information-technology Promotion Agency (IPA).

REFERENCES

[Berkel93c] Kees van Berkel. Handshake Circuits: an Asynchronous Architecture for VLSI Programming. Volume 5 of International Series on Parallel Computation. Cambridge University Press, 1993.

[Blunno98] Ivan Blunno. VL2ASTG Translator from Verilog to Signal Transition Graph, April 1998. User's manual in http://polimage.polito.it/~blunno/vl2astg.html.

[Chu87] Tam-Anh Chu. Synthesis of Self-Timed VLSI Circuits from Graph-Theoretic Specifications. Ph.D. thesis, MIT Laboratory for Computer Science, June 1987.

[Cortadella96b] J. Cortadella, M. Kishinevsky, A. Kondratyev, L. Lavagno, A. Yakovlev. Petrify: a tool for manipulating concurrent specifications and synthesis of asynchronous controllers. In XI Conference on Design of Integrated Circuits and Systems, Barcelona, November 1996.

[Cortadella98] Jordi Cortadella. Petrify Manual Page. Department of Software, Universitat Politecnica de Catalunya, Barcelona, Spain. Manual in the software distribution http://www.lsi.upc.es/~jordic/petrify/petrify.html.

[Furber96] S. B. Furber and J. Liu. Dynamic logic in four-phase micropipelines. In Proc. International Symposium on Advanced Research in Asynchronous Circuits and Systems. IEEE Computer Society Press, March 1996.

[Helbig95] Johannes Helbig. Extending VHDL for state based specifications. In Asia-Pacific Conference on Hardware Description Languages (APCHDL), pages 675-684, 1995.

[Kondratyev99a] A. Kondratyev, J. Cortadella, M. Kishinevsky, L. Lavagno, A. Yakovlev. Automatic synthesis and optimization of partially specified asynchronous systems. In Proc. ACM/IEEE Design Automation Conference, pages 110-115, 1999.

[Kudva95] P. Kudva, G. Gopalakrishnan, and V. Akella. High level synthesis of asynchronous circuit targeting state machine controllers. Asia-Pacific Conference on Hardware Description Languages (APCHDL), pages 605-610, 1995.

[Martin87] Alain J. Martin. A synthesis method for self-timed VLSI circuits. In Proc. International Conf. Computer Design (ICCD), pages 224-229, Rye Brook, NY, 1987. IEEE Computer Society Press.

[Martin97] A. J. Martin, A. Lines, R. Manohar, M. Nystroem, P. Penzes, R. Southworth, and U. Cummings. The design of an asynchronous MIPS R3000 microprocessor. In Advanced Research in VLSI, pages 164-181, September 1997.

[Nowick93a] Steven M. Nowick. Automatic Synthesis of Burst-Mode Asynchronous Controllers. Ph.D. thesis, Stanford University, Department of Computer Science, 1993.

[Plana98] L. A. Plana, S. M. Nowick. Architectural optimization for low-power nonpipelined asynchronous systems. IEEE Trans. on VLSI Systems, 6(1):56-65, March 1998.

[Yun94] Kenneth Yi Yun. Synthesis of Asynchronous Controllers for Heterogeneous Systems. Ph.D. thesis, Stanford University, August 1994.

TOOL PERFORMANCE

Automating the Validation of Hardware Description Language Processing Tools

Sathyanarayanan Seshadri[1], Sanjeev Thiyagarajan[2], John Willis[3], and Gregory D. Peterson[4]
Intel, University of Cincinnati, FTL Systems, and The University of Tennessee

Key words: Automation, Compliance Testing, VHDL, VHDL-AMS

Abstract: Testing the adherence of a hardware description language (HDL) tool to an associated HDL specification is of critical importance to design teams. This paper will develop practical test suite requirements, outline FTL System's VIVA™ technology for automatically generating and applying test suites, and describe the experimental results using VIVA (VHDL Interactive Validation Alchemy) to test FTL Systems' Auriga ® family of HDL compilers.

1.1   Test Requirements

1.1.1   Why Care About Language Conformity?

Design teams make a substantial investment in embodying their designs using HDLs, training designers in a particular HDL, and creating tool suites required to complete a design flow. Preservation of this investment requires

[1] Sathyanarayanan Seshadri (sathyanarayanan.seshadri@intel.com) was with FTL Systems as a student intern and with the University of Cincinnati. He is currently at Intel.
[2] Sanjeev Thiyagarajan (sthiyaga@ececs.uc.edu) was with FTL Systems as a student intern and is currently with the University of Cincinnati.
[3] John Willis (jwillis@ftlsys.com) is the President and CEO of FTL Systems.
[4] Gregory D. Peterson (gdp@utk.edu) was with FTL Systems and is currently at The University of Tennessee.

P.J. Ashenden et al. (eds.), System-on-Chip Methodologies & Design Languages, 233–244.
© 2001 *Kluwer Academic Publishers. Printed in the Netherlands.*

that all tools in the flow conform, within a documented subset, to a known HDL specification.

The development of vendor -neutral languages, such as VHDL and Verilog, provides the common specification. But how do design teams and tool developers insure the adherence of their HDL tools to the specification? This paper establishes metrication requirements, and then describes one technology for meeting these requirements: FTL Systems' VIVA [1].

1.1.2 White-Box and Black-Box Methods

Approaches to software testing are generally divided into white-box and black-box methods. White-box testing uses knowledge and observation of a program's internal functions to ascertain that test objectives are met. In contrast, black-box testing uses the choice of program inputs and observed program response to confirm that objectives are met.

White-box testing can normally only be done by the program developers or a closely related team -- not by an end-user of the HDL tool. This testing method is also further removed from the inputs and outputs that are directly observable by a design team. Because of these limitations, this paper will be exclusively concerned with the creation and use of black-box testing methods for HDL tools. For further information on white-box testing, the interested reader is referred to [2,3].

1.1.3 How Many Test Cases are Needed?

The number and distinctness of test cases in the test suite directly determines the effectiveness of black- box testing. For HDLs such as VHDL and Verilog, the number of tests required to completely test the adherence of a tool to the specification is theoretically unbounded and the test generation problem is NP-complete [4]. Practical black-box testing depends on determining a feasible number of distinct tests. We will consider three methods to establish a practical size for the test suite pool.

1.1.4 Language Test Point Criteria

Using funding from the Air Force Research Laboratory, several efforts have used manual test case generation to produce language conformity test suites. Efforts by MCC (VHDL Version 7.2 updated to 1076-87), Intermetrics (VHDL-87), and the VHDL Technology Group (VHDL-87 and limited VHDL- 93) resulted in from 300 to 3,200 test cases.

The largest of these efforts, by The VHDL Technology Group [5], used function-point analysis of the VHDL language reference manual to produce

3,200 analysis-time test cases. The process involved approximately 10 person-years of effort. The authors of this work estimate that completing the analysis-time test case coverage of VHDL would require an additional 10,000 test cases.

Extrapolating the VHDL Technology Group's effort to embrace VHDL-related standards now in use or standardization suggests that 20,000 to 120,000 test cases are needed. The additional functionality tests VHDL-AMS, shared variables, postponed processes, static elaboration, dynamic elaboration, VHDL-PLI, and execution of both correct and erroneous VHDL code.

Based on the assumption that a trained VHDL programmer can produce one distinct test case per hour, 10 to 40 person-years of effort would be required every five years to update validation tests. Such an effort would require an ongoing multi-million dollar budget and several dozen skilled VHDL programmers, test-case designers, and support staff.

Doubling the cost of VHDL test case development to cover Verilog-related standards, it becomes apparent that the cost of such manual test case development is difficult to justify financially (and to staff in a tight labor market).

1.1.5   Available Test Case Criteria

For its Verilog-based simulators, Cadence Design Systems has taken the alternative approach of collecting end-user test cases for validation testing. According to their presentations at the 1999 Design Automation Conference, Cadence uses approximately 20K user test cases for validation of its digital Verilog simulator products. This number correlates well with the lower bound established above, however it is unclear how distinct a random sampling of designer test cases would be, nor how readily the orthogonality of such test cases can be determined.

1.1.6   Implementation Complexity Criteria

For its Auriga ® line of scalable compilers, FTL Systems' standard for product release requires approximately ten lines of test suite code per line of compiler and runtime code in the tool for each distinct source language. Furthermore, tests must be sufficiently distinct to create a spanning test of language functionality, including both correct and incorrect usage.

Compiler products such as those in FTL Systems' Auriga ® family include approximately one to two million lines of code. The above criteria implies a need for at least 10 to 20 million lines of distinct test suite code. At an average of ten lines of code per test, 1 to 2 million test cases would be needed. Note that thousands of lines in common setup and monitoring code is not included in this estimate. Even if each test included 100 lines of unique code (unusual), no less than 100K independent test cases would be needed to effectively measure the adherence of a digital and mixed-signal tool to its specification.

At an average of 100 dollars/Euro per test case (including overhead), a 10M to 100M dollar budget would be needed for such large-scale manual test case development. Of course such an investment is not financially feasible for a 1B to 2B dollar/Euro total annual market.

Furthermore, the time needed to sequentially compile and execute 1M test cases for regression on each version of a compiler family is also infeasible. Industry requirements dictate validation of each of eight product configurations on more than a dozen platform configurations. Even with a one second compile and execute cycle for each test case (aggressive by industry standards), a single computer could apply no more than about 50K test cases per day. At this rate, 1M test cases would require approximately 20 days for each configuration. When supporting mega-projects with 1M dollar/Euro delay penalties per day, regression turn-around within less than 24 hours is essential.

By 1996, it was clear to the Auriga ® development team that a radically new approach to developing test cases was needed and VIVA™ (VHDL Interactive Validation Alchemy) was conceived. VIVA would semi-automatically generate and maintain over 1M orthogonal test cases for each supported language while consuming the skills of no more than one to two people.

1.2   General Approach

VIVA's generation technology derives an *asymptotically spanning* and *pruned* set of compile and run- time test cases from the formal specification for an HDL. "Asymptotically spanning" means that the longer the test case generator and validation process runs, the closer the validation process will come to the theoretically unreachable goal of complete validation. Furthermore, the progression of tests must be such that most language conformity errors are detected early in the test sequence and any detected failures result in a pruning of the test generation sequence. More information on the general approach can be found in [6].

Figure 1 illustrates VIVA's overall test case generation strategy. A formal specification of the HDL must initially be prepared by manual derivation from the HDL specification. Analysis of the formal specification compiles into a set of C++ classes representing the HDL including:

Lexical

Syntactic
Analysis-time semantics
Elaboration-time semantics
Liveliness
Functional operation
Temporal operation

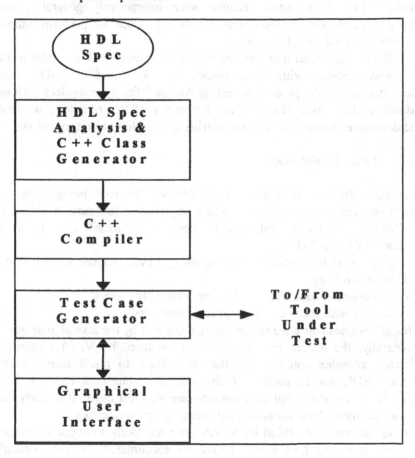

Figure 1. VIVA Conceptual Approach

A conventional, platform-specific C++ compiler translates the classes resulting from analysis of the formal HDL specification into an executable test case generator. This generator interfaces with a graphical user interface to provide runtime control over the testing process. Section 4.0 describes the process of applying these tests to a tool.

Generation of lexical, syntactic, and semantic validation tests are tightly interrelated. While a test might focus on a particular lexical feature, it must be syntactically and semantically valid in all other respects. The need to enforce test point orthogonality presented VIVA's™ designers with substantial but essential design complexity.

Furthermore, a test case intended to generate an error must not accidentally be legal for completely different reasons. For example, a lexeme intended to trigger an error when recognizing an incorrectly formed decimal literal must not inadvertently form an identifier with legally valid meaning. This is essential because when interactively generating and applying several million test cases, it is infeasible to perform human interpretation of individual test cases.

Finally, it is essential that tests be classified as correct or incorrect to the HDL specification without reference to a "golden" HDL tool implementation. In the process of testing Auriga ® family compilers with the thousands of test cases already used for earlier HDL compilers, we have found deviations in every one of these earlier tools. No golden tool exists.

1.2.1   Lexical Validation

Lexical validation measures the ability of the tool being tested to properly segment source code files into a sequence of lexemes. Unlike Lex and YACC, VIVA's lexical specification is integrated into a single specification file (vhdl.viv).

As a result of four years of testing using VIVA, lexical validation has found errors involving:
*   VHDL extended identifiers containing format characters
*   Comments with specific non-graphical characters
*   Illegal lexemes that were not properly processed by the lexical analyzer

Generally, the lexical processing errors identified by VIVA required a moderately complex context, one that is unlikely to result from human-generated HDL source code. If the Auriga ® compilers had not been previously run against about 20K human-generated test cases, it is likely that VIVA would have identified additional lexical processing errors.

While the errors identified by VIVA were not likely to occur in human-generated code, the tool errors could be encountered by automatically generated VHDL code, especially code using characters beyond the Latin-I code page. Although not currently compliant with the VHDL language specification, these additional characters are often encountered in practice.

1.2.2   Syntactic Validation

Syntactic validation evaluates a tool's ability to either correctly parse lexeme sequences into a valid production or to briefly explain why a lexeme sequence lacks syntactic validity. Generally, syntactic productions must reconcile strategy imposed from above in the production hierarchy with semantic constraints propagated from child productions.

An HDL analyzer examines current and potential look-ahead lexemes in order to determine that either no production, a unique production, or multiple productions match the lexeme sequence. Conversely, the VIVA™ test case generator must choose among potential multiple subsequent lexemes to produce a legal (or intentionally illegal) test case.

Most HDLs require that a symbol be defined before its first use. Thus a declaration must be entered into VIVA's output stream before it can be referenced. Such dependencies can be implemented by either "back-patching" the HDL source code stream or formulating top-down strategies that insure the required declarations occur before their first use. Back-patching generally requires arbitrary buffering of the output stream with suitable structure. Top-down strategy requires completely predicting context in advance of use (for example, the occurrence of types before declarations before expressions). VIVA uses both back-patching and top- down or forward-looking strategy techniques for test case generation (back-patching provides a back-up to the forward-looking strategy).

Use of VIVA for syntactic validation of Auriga® family compilers has identified several tool flaws that were unlikely to occur in actual designer code, but that were nevertheless legal. For example, resolution functions may appear in the subtype indication of a case statement' s choice (even though they have no meaning in this context). At least one form of subtype indication was not correctly handled with a resolution function.

1.2.3   Analysis-Time Semantic Validation

Analysis-time validation evaluates a tool's ability to correctly handle semantic features of the language such as visibility, overload resolution, type/nature compatibility, and comparable functionality. Earlier parser technology typically relied on embedded program fragments to implement syntactic functionality.

VIVA operation is facilitated by merging semantic specifications into the grammar files in such a way that the VIVA C++ code generator can analyze semantic constraints and use these constraints within the generation strategy. Given the complexity of VHDL's visibility and overloading rules, formal specification requires that productions have associated visibility regions and visibility rules.

VIVA's™ greatest value in semantic testing occurs in the context of VHDL-AMS and Verilog-AMS, where the actual semantics related to natures have some initial definition ambiguities. In several cases the VHDL-

AMS specification has proven inconsistent with what a designer might expect.

Semantic constraints are intrinsically more diffuse in the language specification than either lexical or syntactic aspects. A semantic constraint or rule may impact many different, unrelated productions. Having a formal specification methodology greatly facilitates such function point interaction compared to earlier test case generation techniques (based on manual function point analysis). Previously, test case designers had to respond to language changes by manually spotting interactions.

1.2.4 Elaboration-Time Semantic Validation

Elaboration-time validation evaluates a tool's ability to perform both static (structural) and dynamic (subprogram) elaboration. While tests are initially generated as source code corresponding to analysis units, tests generated for elaboration must correctly predict properties of the binding scopes present after elaboration.

The test case generator must retain the state associated with each elaborated instance of a design unit (or module) and track constraints/strategy on an instance-by-instance basis. Since the coupling of instances through component instantiation and function call sites does not occur until after the structure and subprograms have been defined, a mixture of back-patching and top-down strategy must be used to insure that the resulting test case correctly embodies the required test objectives.

Recursive configurations and subprograms are a particularly interesting aspect of elaboration-time semantic validation. Many forms of structural and subprogram recursion can only be bounded during runtime, forcing VIVA to track elaborated state, strategy, and constraints symbolically rather than by enumeration via the use of context-based constraint evaluation [6].

1.2.5 Liveliness Validation

Liveliness validation establishes a tool's ability to interact with its environment through assert statements, report statements, Text I/O, and programming language interfaces. Given the substantial flexibility of such interfaces, VIVA probes liveliness via parametric permutation of test case templates. Permutations include data types/natures, input data sequences, and output sequences.

1.2.6   Functional Operation

Functional validation evaluates a tool's ability to compute expressions and assign sequential side-effects with partial orderings compliant to the language specifications. In order to de-couple distinct functional tests, it is useful to encapsulate each test in a distinct process (subsequent, temporal tests explore interaction between processes and between simple simultaneous equations).

The greatest challenge presented by functional verification results from predicting the results of operations without reliance on a golden tool. One way of handling this computation is the formal construction of alternative operator sequences that logically result in the same output. Outputs can then be compared.

Functional verification has been instrumental in identifying platform-specific code generation errors (code generators incorrectly emitting instructions, generally in the presence of data type boundary conditions). Hardware description languages traditionally do not have well-defined semantics in the case of exception conditions.

1.2.7   Temporal Operation

Temporal validation provides the final step, insuring that processes side-effect signals and variables in a correct order and that analog quantities evolve with a trace history consistent with the HDL specification. VHDL shared variables, Verilog nets, and analog quantities cannot be specified a-priori. Rather, the test case generator must determine if the resultant value belongs to the set of acceptable values. Temporal validation is the most complex part of VIVA^{TM} as well as potentially the most valuable in terms of the number of subtle tool errors it detects.

1.3   Testing Approach

VIVA's automatic test case generation capability can generate tens to hundreds of test cases per second. Most VHDL tools require at least a second to compile and execute a test case. We have seen tools from other groups, often using intermediate compilation stages, that require more than a minute to compile even small test cases.

Given the execution bottleneck, the need for concurrent test case execution was seen early in FTL Systems' development process. This realization lead to the parallel runtime structure shown in Figure 2. In this structure, a common graphical user interface controls one or more test case

242

generators. Each test case generator drives one or more tools under test via a wrapper.

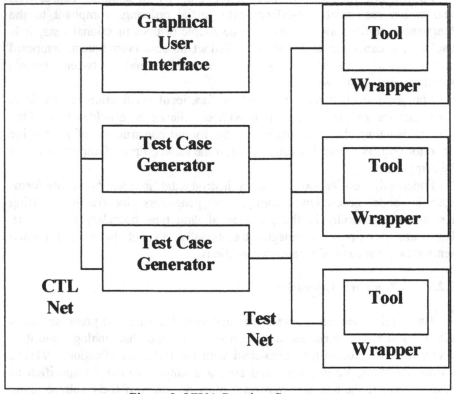

Figure 2. VIVA Runtime Structure

Wrappers surrounding each tool, written in Perl, allow rapid modification for different tool interfaces and different response sequences from the tool being tested. Perl's internal network support allows easy interface to the TCP/IP-based test network.

The tool being tested can include a compiler/simulator, synthesis/simulator, emulator, compiler/processor, formal verification tool, or any combination of the above. All that the tool needs to provide is an embodiment of an arbitrary HDL design with accessible input and output through which to probe the tool's behavior or predicates.

Each test case generator is responsible for tracking one or more pending test case executions. When the test case results return from execution, the test case generator abstracts the test results for reporting on the CTL network. Other test case generators may use this information to prune test case generation or to skew probability distributions employed in randomized test generation schemes [6]. The graphical user interface summarizes test case results in terms of the original, formal language specification.

1.4   Conclusions

Over the last five years, VIVA^{TM} has played and continues to play an essential role in the development of high-fidelity, optimizing compilers at FTL Systems. The technology provides a nearly spanning set of test cases for the development and regression testing of VHDL and other language implementations. Its development cost and ongoing maintenance is generally affordable in the context of major commercial compiler development.

By starting from a terse formal language specification, VIVA is much more adaptable to evolving source language requirements. As a result of VIVA, initial releases of Auriga ® compilers benefit from running a wider variety and number of test cases than many compilers encounter during their life cycle.

1.5   Acknowledgments

Engineers Robert Newshutz (early VIVA compiler) and Patrick Gallagher (runtime environment and graphical user interface) also worked on the development of VIVA at FTL Systems. At subcontractor Clifton Laboratory, Phil Wilsey lead a team including Dale Martin and Darryl Dieckman. Thanks are due to AFRL Program Manager Steve Drager as well as DARPA Program Managers Robert Parker and Jose Munoz. Thanks also go to Hal Carter of the University of Cincinnati for acting as the thesis advisor for Sathya and Sanjeev.

This work was funded by the Air Force Research Laboratory under Contract F33615-96-C-1909 (digital VHDL) and Contract F30602-98-C-0046 as well as DARPA Contract DABT63-96-C-0004 and Contract DABT63-97-C-0038. Subsequently, VIVA has been commercially funded by FTL Systems. Clifton Laboratory was a subcontractor to FTL Systems for some digital VHDL-related portions of the above work. This paper does not necessarily reflect the position or policy of the US Government. VIVA^{TM} and Auriga ® are trademarks or registered trademarks of FTL Systems, Inc., all rights reserved. VIVA is patent-pending internationally. VIVA is currently a technology demonstration/validation and not an announced commercial product. For further information on VIVA, write ftlsys@ftlsys.com.

1.6   References

[1] John Willis, Philip A. Wilsey, Gregory D. Peterson, John Hines, and William Dashiell, "Semi-Automatic Validation of VHDL & Related Languages." In the *Proceedings of the Fall 1996 VHDL International Users' Forum.*

[2] Robert M. Poston, *Automating Specification-Based Software Testing*, IEEE Press, 1996.

[3] Paul C. Jorgensen, *Software Testing, A Craftman's Approach*, CRC Press, 1995.

[4] John D. Musa and Franck Ackerman, "Quantifying Software Validation: When to Stop Testing?," *IEEE Software*: 6(3), May, 1989.

[5] John W . Hines and Bill Billowitch, "Development of a VHDL Validation Suite."   In the *Proceedings of the Spring 1995 VHDL International User' s Conference.*

[6] Sathyanarayanan Seshadri, *An Approach to Automatic Test Generation for Evaluating VHDL-AMS Simulators*, Master of Science Thesis, University of Cincinnati, July, 2000.

A RETARGETABLE SOFTWARE POWER ESTIMATION METHODOLOGY

Carlo Brandolese

Politecnico di Milano - DEI

P.zza L. da Vinci, 32 - 20133 Milano, Italy

brandole@elet.polimi.it

Abstract   In the design of mixed hardware/software embedded systems, the early assessment of the power budget is a key factor to concurrently meet time-to-market and product competitiveness. An increasing contribution to the overall power consumption depends on the software portion of the design and is influenced both by the system specification style and by the target microprocessor. The proposed estimation methodology operates at an abstraction level halfway between the system language and the specific assembly language and provides power consumption figures useful for both source code analysis and modification, and for target processor selection.

1.   INTRODUCTION

The importance of the power constraints in the design of embedded systems has continuously increased in the past years, due to technological trends toward high-level integration and increasing operating frequencies, combined with the growing demand of portable systems. So far, only a few co-design approaches have considered power consumption as a comprehensive system-level metric [1] [2] [3]. According to [4], the methods to estimate the software power consumption can be grouped in three classes: gate-level processor simulation [5], architectural-level processor description and instruction-level models. The gate-level simulation provides the most accurate results at the cost of extremely time demanding simulations. Furthermore, due to the lack of information of the processor gate-level description, this methodology is rarely viable. Architectural-level power estimation is less precise but much faster than

245

P.J. Ashenden et al. (eds.), System-on-Chip Methodologies & Design Languages, 245–254.
© 2001 *Kluwer Academic Publishers. Printed in the Netherlands.*

gate-level estimation [6] and requires a coarser grain model of the processor (ALU, register file, etc.).

Instruction-level power estimates are typically based on stochastic data modeling of the current drawn by the processor for each instruction. Such methodologies have been proposed in [2] and [3]. The goal of this paper is to describe a power estimation methodology for software components suitable for typical hardware/software embedded system architectures. The proposed approach allows exploring the design space either to early retarget architectural design choices, or to redesign the power-critical parts of the system. The proposed technique has been implemented in the TOSCA co-design framework for control-dominated embedded systems [10]. The description language adopted in the framework is OCCAM2, since it allows describing mixed hardware/software systems, and provides constructs for parallel execution and process synchronization with the rendez-vous semantic. It is worth noting that the use of OCCAM2 as description language, does not affect the generality of the proposed approach, the key point being the introduction of an intermediate pseudo-assembly language, called VIS (*Virtual Instruction Set* [7]).

The VIS model provides sufficient detail to permit an accurate estimate and a degree of generality allowing easy retargeting towards different processors. The paper is organized as follows. In section 2, the TOSCA framework, and in particular software flow, are described; in section 3, the procedure adopted to derive the power estimation model is detailed and justified; in section 4, the results obtained on some benchmarks and an industrial design are reported; finally, in section 5 conclusions are drawn and an outline of the current research is given.

2. FRAMEWORK AND FLOW

When designing low-power embedded systems, and, in particular, their software components, two aspects have to be considered. On one hand, the specification can be partitioned between hardware and software in different ways; on the other hand, the same source code results in different power requirements when compiled and executed on different microprocessors. The designer, usually, performs the partitioning on the basis of figures such as cost and performance. With the increasing demand of portable devices, considering the power consumption from the early phases of the design, becomes essential. A satisfactory design-space exploration should consider these aspects concurrently and should be fast enough not to compromise time-to-market constraints.

As far as the software portion of the design is concerned, the analysis of different solutions would require a complete development environment (compiler, debugger, instruction-set-simulator, etc.) for each target microprocessor considered. In addition, a modification of the source code implies re-running the complete flow, from compilation, through profiling and power estimation, to back-annotation. Currently only a few microprocessor vendors ship such a complete development toolkit. This means that it is often necessary to complete (or, at least, customize) the vendor toolkits with in-house developed tools and utilities. The TOSCA environment overcomes these problems with the introduction of the intermediate, pseudo-assembly VIS. This language is sufficiently close to an actual assembly to guarantee a good degree of detail, and, at the same time, sufficiently general to permit fast retargeting towards different assembly languages.

The proposed power estimation methodology operates at the VIS level, and achieves a good trade-off between accuracy and effectiveness. Figure 1.1 depicts the part of the TOSCA framework devoted to the software flow. In this figure, rounded boxes indicate input and output files while grayed-out boxes indicate tools. The software power estimation flow is composed of five tools:

- **Compiler** - An OCCAM2 to VIS compiler.

- **Simulator** - A VIS simulator and profiler.

- **Mapper** - A library-based translator from VIS to different real target assembly languages.

- **Estimator** - A library-based power estimator.

- **Back-annotator** - A tool for assembly to VIS and VIS to OC-CAM2 back-annotation.

The evaluation of power consumption of a given OCCAM2 source code begins within the compilation in VIS code. The VIS code can then be debugged and validated, executing it on the virtual *VIS Machine* implemented in the simulator. When the functionality of the VIS code has proven correct, the simulator can be run in batch mode to obtain profiling information, i.e. the number of times that each VIS instruction has been executed. The VIS code is then *mapped* to a specific, commercial assembly language, according to a fixed set of *rules* and its power consumption is evaluated on the basis of power figures collected into *annotation libraries*.

Exploring different solutions targeted to different microprocessors would require re-mapping and re-estimating each and every alternative. This

248

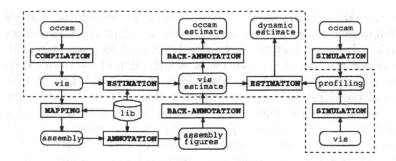

Figure 1.1.   The TOSCA software flow

can be avoided noting that the same VIS instruction (same op-code, same operands type and value) is always mapped to the same assembly portion of code. This knowledge, along with an accurate statistical analysis of the results (see sections 3 and 4), allows characterizing directly each VIS instruction in terms of power consumption, once the target microprocessor is selected. This approach shortens the design flow, since avoids the mapping, the assembly annotation and one of the back-annotation steps (see figure 1.1). Furthermore, when a dynamic analysis is desired, a single session of profiling at VIS level is necessary.

It is worth noting that, being each VIS instruction power-characterized for different architectures, the power consumption impact of the choice of a specific target microprocessor can be estimated by simply switching to the suitable annotation library, without simulating or mapping. Furthermore, the static or dynamic power consumption figures gathered at VIS level can be conveniently back-annotated to the OCCAM2 source code, giving a clearer and more useful view of the overall power budget of the software components. Typically, the back-annotated information are valuable hints to the designer on what parts of the source code should possibly be modified or moved to the hardware partition.

3.   METHODOLOGY

The methodology presented in this section can be applied to derive the power characterization of the VIS instruction set for any target processor. A VIS instruction is defined by:

- *op-code:* the type of operation to perform;

- *addressing mode:* the type of operand used;

- *operand value:* the value of the operands;

The *class* of an instruction is defined by its op-code and the addressing mode of its operand, but ignoring the value of operands. For each instruction *class*, thus, a set of instructions can be built, varying the value of the operands.

As an exemple, consider the VIS instruction MOVE.W #16, +5(R0): the op-code is MOVE.W, the addressing mode of the two operands are *immediate* (#16) and *indirect* (+5(R0)) and the values are 16 for the first operand and the couple (5, R0) for the second. It is not uncommon that instructions of the same *class* map to different assembly codes and are thus characterized by different power consumption [9] [11].

As an example consider the ARM7TDMI microprocessor: an immediate constant can be loaded into a 32-bit register directly if and only if it falls in the range 0–255; when the immediate value is greater than 255, its low and high bytes must be loaded into the register separately, suitably shifting the register content after the first load. To properly account for these differences *all* the possible instructions of a given instruction *class* must be analyzed.

Let $\mathcal{I}$ be an instruction *class* and $i_j \in \mathcal{I}$ a generic instruction with specific operands values. Using the estimation flow described in section 2, all instructions $i_j \in \mathcal{I}$ can be annotated with the actual timing $t_j = t(i_j)$ and average current $c_j = c(i_j)$ leading to two vectors of measures $\mathbf{T} = \{t_j\}$ and $\mathbf{C} = \{c_j\}$.

To derive single values $t(\mathcal{I})$ and $c(\mathcal{I})$ for the VIS instruction *class* $\mathcal{I}$, four different approaches have been adopted:

- **Different translations only.** The timing $t(\mathcal{I})$ and current $c(\mathcal{I})$ are computed by averaging only the values t_j and c_j corresponding to different translations. This method is easily applicable because the mapping *rules* are known and thus it is possible, for each instruction *class*, to build all the instructions necessary to exercise all the cases that the specific *rule* considers. This strategy is supported by the hypothesis that a generic VIS code exploits in a statistically uniform way all the possible alternatives of all the *rules*.

- **Different figures only.** The timing and current are computed by averaging only the values t_j and c_j that are different. This approach is justified by the consideration that different translations may lead to identical total power figures.

- **Complete uniform.** The timing and current are computed averaging *all* the values t_j and c_j. This choice is statistically justified considering that, when a sufficiently long VIS program is considered, and the semantics of the instructions is neglected, the distribution of instructions within an instruction *class* tends to be uniform. This is mostly due to the fact that there is no reason to suppose non-uniform the distribution of values of the operands.

- **Complete weighted.** The timing and current are computed by averaging *all* the values t_j and c_j, *weighted* with relative frequencies obtained from an analysis of a sufficiently large set of benchmarks. The previous case, as mentioned, ignores the semantics of the instructions. This hypothesis can be removed considering the fact that some values are more likely to be used than others. The measured relative frequency is thus considered an estimate of the probability.

Once all VIS instructions have been characterized, according to one of these strategies, in terms of number of clock cycles and average current consumption, power estimates of a generic VIS program can be derived. The estimation flow produces, for each instruction i_k of the VIS program, the number of clock cycles required for execution t_k, the average current absorbed per clock cycle c_k and the number of times n_k the instruction has been executed during a simulation session driven by typical input streams. The energy absorbed by all executions of the k-th instruction is $e_j = n_k V_{dd} t_k c_k$ and the total energy and cycles are:

$$E \;=\; \sum_{k=1}^{M} e_k = V_{dd} \sum_{k=1}^{M} n_k t_k c_k \tag{1}$$

$$T \;=\; \sum_{k=1}^{M} n_k t_k \tag{2}$$

where V_{dd} is the power supply of the core and M is the number of instructions of the considered VIS program. The average power consumption is thus:

$$\overline{W} = \frac{E}{T} = V_{dd} \cdot \frac{\sum_{k=1}^{M} n_k t_k c_k}{\sum_{k=1}^{M} n_k t_k} \tag{3}$$

In the next section the four proposed characterization approaches are compared.

4. EXPERIMENTAL RESULTS

4.1. METHODOLOGY TUNING

The four modeling approaches presented in the previous section have been applied to the entire VIS instruction set for the two commercial processors ARM7TDMI in Thumb mode (ARM Ltd.) and MC68000 (Motorola) and the results for the first are reported in figure 1.2 (similar results have been obtained for the Motorola processor, but are omitted here for brevity). The figure reports relative errors with respect to the *complete uniform* method, and shows that the four methods, for each instruction *class* lead to significantly different figures (up to ±50%).

To select the most accurate method, benchmarking is necessary. Some sample OCCAM2 sources have been run through the complete compilation, mapping, power annotation and back-annotation flow and the power measures obtained have been compared against the four estimation methods, leading to the results summarized in table 1.1. Note that these errors are much smaller than those plotted in figure 1.2: this is due to cancellation effects that occur over real VIS instruction sequences.

Table 1.1. Relative errors (absolute values)

Method	Error
Different figures only	7.21%
Different translations only	4.31%
Complete uniform	3.14%
Complete weighted	2.71%

The comparison of the results shows that the two best methods are the *complete uniform* and the *complete weighted*, the latter being slightly more accurate. In the next paragraph the methodology is applied to an industrial example and results are discussed.

4.2. TEST CASE

To validate the presented methodology and framework, a set of benchmarks has been built. This set comprises OCCAM2 programs 10 to 230 lines long. The result of compilation of these codes are VIS files with lengths ranging from 60 to 2,600 lines, approximately. The benchmarks have been used: to decide for one of the strategies for VIS instruction set characterization (see table 1.1); to measure the accuracy of the VIS-level model and to measure the speed-up obtained with the estimation flow with respect to the complete flow. Figure 1.3 reports the run times of the

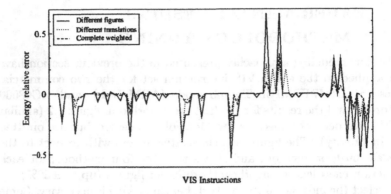

Figure 1.2. Energy errors relative to the *complete uniform* method

complete flow and the estimation flow on the benchmarks. The speed-up varies from 30% to 75% with an average value of 63%, confirming the suitability of the proposed methodology as an effective design-space exploration strategy.

To determine the accuracy of the proposed model, the methodology has been applied to a real industrial design called ILC16, a 16-channel link controller developed at Italtel R&D Labs [12], and the results, summarized in tables 1.2 and 1.3, have been obtained.

Table 1.2 . Run times for the ILC16 application

Processor	Measure	Estimation	Speed-up
ARM7TDMI	37.57 s	22.07 s	58.8%
MC60000	53.17 s	29.96 s	56.3%

Table 1.3 . Power estimates for ILC16

Processor	Measure	Model	Error
ARM7TDMI	224.48 mW	230.65 mW	2.77%
MC68000	26.20 mW	27.41 mW	4.47%

The ILC16 design is composed of 54 OCCAM2 procedures, for an overall line count of 2200 lines. The VIS code resulting from compilation has 55000 lines and the target assembly codes for the ARM and Motorola processors are 75000 and 89000 lines long, respectively.

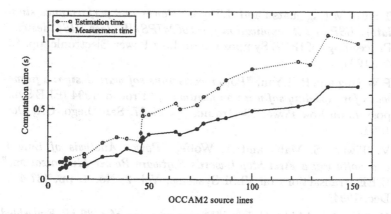

Figure 1.3. Run times of measure and estimation flows

It is worth noting that the average currents for the ARM7TDMI processor—used to characterize the VIS instruction set—have been obtained from actual measurements [11], while those for the MC68000 processor are relative to the power consumption of a reference instruction [8].

5. CONCLUSIONS

The paper presented a new methodology for embedded software power estimation. The main focus of the proposed approach is fast design-space exploration with respect to different coding styles and different target processors. To validate the methodology, an industrial example, used as a test vehicle during two ESPRIT projects, has been analyzed using an integrated environment developed to this purpose within the TOSCA co-design framework. The techniques described, though currently based on the OCCAM2 system modeling formalism, are general and applicable to different high-level languages such as C/C++. The results obtained are encouraging under both accuracy and effectiveness points of view. The current effort of the research is aimed at properly considering dynamic effects such as the presence of a memory hierarchy and pipelining.

References

[1] E. Macii, M. Pedram, F. Somenzi, *"High-Level Power Modeling, Estimation, and Optimization,"* IEEE Transactions on Computer-Aided Design of Integrated Circuits and Systems, Vol. 17, No. 11, 1998.

[2] T.Sato, M.Nagamatsu and H.Tago, *"Power and performance simulator: ESP and its application for 100MIPS/W class RISC design,"* Proceedings of IEEE Symposium on Low Power Electronic, pp. 46-47, 1994.

[3] P.W.Ong and R.H.Yan, *"Power-conscious software design: a framework for modeling software on hardware,"* Proc. of 1994 IEEE Symposium on Low Power Electronic, pp. 36-37, San Diego, CA, Oct. 1994.

[4] V. Tiwari, S. Malik and A. Wolfe, *"Power Analysis of Embedded Software: a First Step towards Software Power Minimization,"* IEEE Transactions on VLSI Systems, Vol. 2, No. 4, pp. 437-445, Dec. 1994.

[5] V. Tiwari and M.T.-C. Lee, *"Power analysis of a 32-bit Embedded Microcontroller,"* VLSI Design Journal, 1996.

[6] J.Russell, M.F.Jacome, *"Software Power Estimation and Optimization for High Performance, 32-bit Embedded Processors,"* Proc. of ICCD'98, International Conference on Computer Design, Austin, Texas, USA, October, 1998.

[7] C. Brandolese, W. Fornaciari, F. Salice, D. Sciuto *"Fast Software-Level Power Estimation for Design Space Exploration,"* Politecnico di Milano, Tech. Report 99.62, 1999.

[8] C. Brandolese, W. Fornaciari, F. Salice, D. Sciuto *"An Energy Estimation Model for 32-bit Microprocessors,"* Politecnico di Milano, Tech. Report 99.63, 1999.

[9] PEOPLE ESPRIT project n.26769, Deliverable D1.3.2.

[10] PEOPLE ESPRIT project n.26769, Deliverable D1.3.3.

[11] PEOPLE ESPRIT project n.26769, Deliverable D1.2.1.

[12] A.Allara, M.Bombana, W. Fornaciari, F.Salice, *"A Case Study in Design Space Exploration: The TOSCA Environment Applied to a Telecom Link Controller,"* IEEE Design & Test of Computers, 2000, (to appear).

Performance Tradeoffs for Emulation, Hardware Acceleration, and Simulation

Gregory D. Peterson

Electrical and Computer Engineering

The University of Tennessee

(865)-974-6352

gdp@utk.edu

Key words: Performance evaluation, verification, emulation, hardware acceleration, simulation, tradeoffs

Abstract: *Performance demands for design verification continue to grow (exponentially) with the size of designs. Consequently, the amount of time spent verifying a particular design meets its specification is taking an ever-increasing proportion of the design cycle time. Given the substantial costs and increasing importance of system verification technologies, determining the best verification strategy is critical to design and business success. This paper focuses on how a designer can develop and use a performance/cost model to perform tradeoffs between the different types of emulation, hardware acceleration, and simulation tools available. Different forms of model and parameters are explored, with some typical engineering examples illustrated.*

1.1   Introduction

Moore's Law continues to hold as electronic designs double their potential size every eighteen months. System on a Chip (SoC) designs comprised of tens of millions of gates with clock speeds approaching 1 GHz are in development and will become commonplace over the next few years. Because of the shrinking times to market, high costs in fabrication, and increasing circuit complexity, the verification task is now taking 50-80% of the overall design effort [3]. Performance demands for design verification continue to grow (exponentially) with the size of designs. With the market

P.J. Ashenden et al. (eds.), System-on-Chip Methodologies & Design Languages, 255–267.

pressures to reduce cycle time exacerbating this situation, the verification needs of designers looms as a significant challenge.

A number of approaches exist for verifying a design before production; popular approaches include physical prototyping, emulation, hardware acceleration, simulation, and formal verification. Particularly for ASICs, physical prototyping is at best an expensive approach to be saved for final sign-off and not an economically effective verification approach for catching most errors. Similarly, formal verification promises the ability to prove that a given design meets its specification (modulo the fidelity of the design representation in describing physical phenomena), but practical limitations concerning the mathematical complexity of formal verification approaches, limits on the size of design that can be processes, and the verification runtime impair this approach. Hence, although there is certainly a place in design flows for prototyping and formal verification, we focus on emulation, hardware acceleration, and simulation for verification in this paper.

Emulation systems use reconfigurable hardware (typically FPGAs) to implement essentially equivalent functionality in hardware. Although subject to constraints on routing and visibility, the designer can expect to see performance within one or two orders of magnitude of the design. To achieve this high performance for test vectors, the design must first be synthesized to the primitive logic elements. Thus, emulators take a significant amount of time up front to complete the mapping of the design onto the primitives. The limited routing available for FPGAs, typically results in lower utilization rates of the reconfigurable hardware. Similarly, limitations on the visibility into an emulated system may necessitate re-mapping of the emulated design onto the reconfigurable hardware to achieve the required visibility.

Hardware acceleration for verification comprises custom hardware that is dedicated to specific simulation applications. The hardware accelerators often use parallel processing approaches to achieve high performance. This approach is similar to cycle simulation because the processors perform the simulation in lock-step for each simulated object at each time point. Thus, for simulations with high activity levels, hardware accelerators can achieve significant performance gains. Given the custom nature of the hardware implementations of simulation hardware accelerators, this approach can be very expensive. With Moore's Law continuing, the depreciation for an investment in this approach is significant as well.

Simulation is the most widely used verification approach and includes event-driven simulation and cycle simulation (among other approaches). Event driven simulation provides the maximum flexibility with respect to timing support and visibility, but often suffers from poor performance with respect to the number of cycles or events processed per second. Compilation of a design for simulation is normally faster than for an emulator or hardware accelerator. Simulation speed is one of the driving factors in technology innovation for design automation and a significant factor is designer productivity. Performance models for serial and parallel simulation algorithms are given, including the effects of other (competing) tasks on the overall simulation performance.

In the following sections of this paper, performance and cost-based models for emulation systems, hardware accelerators, and simulators are developed and compared. Some typical coefficient values are proposed to illustrate the usefulness in making quantitative tradeoffs in determining the most appropriate validation approach for a given design. Different verification strategies are explored to demonstrate how to decide the best approach for a given phase of a design and where a combination of approaches makes sense. A number of design process factors are included in the overall performance model to help optimize verification methodologies. Based on these models and their applications, we conclude with some observations concerning the best way to employ verification techniques.

1.2   Performance Models

Performance models are widely used to determine the most appropriate computer architecture, queue lengths, etc. In developing performance models to help in trading off verification techniques, one goal of this work is to keep the models accurate, but relatively simple and focused on first-order effects. In order to consider the performance expected from various alternative verification technologies, one may begin with a simple description of the time required to complete testing of the model for a number of clock cycles. (See Table 1 for each of the variables used in the models.)

$$T_{\text{VERIFY}} = t_{COMP} + \frac{N_C}{f_{HW}} + N_E \left(t_{RECOMP} + t_{DEBUG} \right) \tag{1}$$

Equation 1 gives a simple expression that illustrates the relationship of the verification time to the frequency of the verification engine and the number of clock cycles simulated. It also includes the time required to compile a model from its source hardware description language to machine code suitable for simulation. Similarly, this model includes the costs associated with debugging the system and recompiling the design. To simplify the model, we assume there are N_E errors that are detected during the verification.

With the appropriate choice of parameter values, this equation can be used as a simple model describing the expected performance of an emulation or hardware acceleration verification engine. Such a simple model can be used because an emulator or hardware accelerator use a straight-forward algorithm for evaluating the value of each gate within the system. Each clock, every gate has its value computed and updated for the next cycle.

Because emulators and hardware accelerators calculate the value of each gate every cycle, they are well suited for post-synthesis regression testing. The frequencies they can achieve approach that of the actual fabricated chip, sometimes coming within an order of magnitude in performance. The inherent focus on gate-level logic verification restricts the applicability of emulation, so verification of systems at higher (or lower) levels of abstraction is generally not possible. Hardware acceleration provides some support for RTL models, but its primary focus is on gate level verification. The limited visibility into the design may necessitate the recompilation or resynthesis of the model to gain the required insight into the behavior within circuit blocks. With these caveats, emulation and hardware acceleration provide a very powerful, though expensive, verification capability to design teams.

The software-based approach to verification, namely simulation, supports a richer variety of timing models, thus enabling verification at different stages of design as well as supporting verification at different fidelities.

For synchronous designs using a subset of HDL constructs, one simple optimization is to use *cycle simulation* as with the hardware verification techniques. This results in a simplified and faster control algorithm for the simulation kernel, at the expense of evaluating potentially extraneous events and restrictions on the timing granularity supported. A simple performance model for a cycle simulator is shown in equation 2, where we use N_C as before, but now must add N_G and t_G to reflect the number of gates to be simulated and the time to complete the simulation computations for each gate, respectively.

$$T_{VERIFY} = t_{COMP} + N_C N_G \, t_G + N_E \left(t_{RECOMP} + t_{DEBUG} \right) \tag{2}$$

In the cycle simulation performance model, we model the compilation time and debugging time as before, but with appropriately chosen parameter values. The time spent per gate completing the necessary simulation computations can be computed for execution on a shared memory symmetric multiprocessor using superscalar processors by applying the following equation. (Note the equation reverts to a single issue uniprocessor computer when N_P and I_i are chosen to be 1.)

$$t_G = \frac{I_G}{N_P \, I_i \, f_{CPU}} \tag{3}$$

Another popular approach to simulation is to model events occurring at discrete points in simulated time, which is known as *discrete event simulation*. Discrete event simulation requires a more sophisticated synchronization mechanism, the event queue, for ensuring the proper causality holds in throughout the course of the simulation. Strictly speaking, the exercise of VHDL and Verilog models is a good example of discrete event simulation; if an appropriate subset of language constructs are used and the reduced timing accuracy acceptable, then such an HDL model can be simulated using cycle simulation techniques. The performance of discrete event simulation is similar to that of cycle simulation, except t_E is used to represent the time spent processing each event and the activity level a is used to represent the proportion of circuit elements which have events to process for any given point in time. The processing time per event t_E can be found as in equation 3, but with I_E substituted in the numerator for I_G. Note that this model is relatively coarse, and assumes that the number of events in the simulation is approximately equal to the product of the activity level and the number of gates. If specific gates have events more often than once per clock period, then a more refined model may be needed.

$$T_{VERIFY} = t_{COMP} + aN_C N_G\ t_E + N_E \left(t_{RECOMP} + t_{DEBUG} \right) \tag{4}$$

In the case of a parallel simulator executing on a distributed memory MIMD configuration, the accurate modeling of its performance requires the consideration of communications, load imbalance, and synchronization costs [8, 9]. In the case of cycle simulation, performance modeling results from the analysis of synchronous iterative, or multiphase, algorithms can be applied [1,8].

$$T_{VERIFY} = t_{COMP} + Nc \left(t_{SER} + \frac{\eta\ t_{PAR}}{N_P} + t_{SYNC} \right) + N_E \left(t_{RECOMP} + t_{DEBUG} \right) \tag{5}$$

A simple but accurate model for the performance of a parallel cycle simulator is given in equation 5. This equation breaks the computation into serial, parallel, and synchronization times per iteration. Depending on the implementation of the simulator, computations such as the model compilation may be either serial or parallel. Ideally, the serial portion of the computation will be minimized to take maximum advantage of the parallel processing resources available. The synchronization time is dependent on the computer architecture used as well as the algorithm used, but should be minimized to improve performance. Finally, the parallel computations are scaled by a load imbalance factor, η. This term can be measured or modeled based on finding the expected maximum drawn from probability distributions for each processor representing the simulation workload, background processing, and differences in processor capabilities in the case of heterogeneous resources being applied to the simulation. See [7,8] for more details about this modeling approach.

In the case of discrete-event simulation, a wide variety of performance evaluation results have been published based on the synchronization protocol used to ensure causality. For more details, see [2,4,6,8,11]. In general, these results are highly dependent on the type of model simulated, the parallel computer architecture used, and the synchronization protocol applied. The designer may use equation 5 for discrete event simulation as well, with appropriate values for the η scale factor.

Upon inspection of the terms in the equations above, the reader may question the means to determine several of the parameter values. Of the parameters above, such values as the compile time, debug time, number of gates, number of cycles simulated, hardware frequencies, and the number of instructions issued per processor clock cycle must be determined by the modeler based on the system modeled or the verification engine being evaluated. Later in the results section, some typical values for these terms are considered when exercising the models listed above..

Perhaps the most controversial parameter that a designer will need to include to use these models is the number of errors in the model. One reason for the difficulty in guessing the expected number of errors stems from overly optimistic designers. In reusing functional blocks from earlier designs or from third parties, estimating the error rate is very difficult. Nonetheless, the earlier errors are identified and fixed, the lower their cost.

If one considers the design of a 5 million gate SoC comprised of a number of 100K gate blocks, and if we assume that each block has a 98% chance of being implemented correctly, then the chance the overall SoC will work correctly can be found using Bernoulli trials (and assuming independent random variables) by the equation 6.

$$\text{Prob[SoC is correct]} = (0.98)^{50} = 0.364 \tag{6}$$

The above example illustrates why system level operation depends on the component blocks being tested sufficiently. Note the above equation also does not consider the interfaces between the blocks, a significant cause of errors. Hence, in practice, 90% of designs work in isolation, but only 50% of systems initially work [3].

The reader is encouraged to estimate the number of errors based on past design efforts. In this paper, the purpose of estimating the number of errors in the system is to help determine how many test vectors will be enough to give sufficient confidence levels and to help make verification strategy decisions based on these estimates.

Table 1. Table 1: Model Parameters

T_{VERIFY}	Verification time
N_G	Number of gates
N_C	Number of cycles/test vectors
N_P	Number of processors
f_{CPU}	Clock frequency of processor
f_{HW}	Frequency of emulator/ hardware accelerator
I_i	Instructions issued per cycle
I_E	Instructions per event
I_G	Instructions per gate simulated
t_E	Time per event
t_G	Time per gate simulated
t_{COMP}	Time to compile/synthesize model
t_{RECOMP}	Time to recompile/resynthesize model
t_{DEBUG}	Time spent debugging each error.
t_{SER}	Time spent on serial computations
t_{PAR}	Time spent on parallel computations
t_{SYNC}	Time spent on synchronization
f_E	Frequency constant of errors detected
N_E	Number of errors
a	Activity level
η	Load imbalance factor
U	Utilization rate

If one assumes that each of the errors is detected in isolation by the test vectors, then a good stochastic model for determining relative verification confidence levels is to assume independent exponential distributions for detecting each of the errors. In this case, the probability of detecting each of the errors can be found with equations 7 and 8 [5].

$$\text{Prob[error detected]} = 1 - e^{-f_E N_C} \qquad (7)$$

$$\text{Prob[all errs detected]} = (1 - e^{-f_E N_C})^{N_e} \qquad (8)$$

Using this equation, an appropriate number of tests can be determined. From a performance perspective for verification regression testing, each detected error will interrupt the testing process while designers seek to identify and fix the error. In the case of very expensive emulation or hardware acceleration equipment, idle time can be quite costly when depreciation is considered. In the case of emulators, hardware accelerators, or simulators on large parallel processors, each error may result in the verification tasks being removed from the batch queue, resulting in potentially long delays before the next available time slot can be used, thus artificially increasing the effective debug/recompile time. Finally, if some amount of recompilation is needed to perform debugging, the expensive verification engines are unavailable to others as well while the errors are identified. Hence the debug time may vary for the different approaches.

1.3   Cost Models

The performance models developed above for comparing the raw performance of difference verification engines can yield great insight into the potential approaches a design team could employ in helping ensure their chips operate correctly.   Few design teams can attack the verification problem without some consideration of the costs involved. Given this fact, we next turn to how best to create pragmatic cost functions to support a quantitative tradeoff between the various verification techniques.

A simple cost function can be created which simply seeks to minimize the verification time, without regard to other costs. In such a case, a design team should simply choose the technology which results in the lowest runtime above.

A variation on this cost function would be to assume there exists a fixed verification budget and to purchase the mix of emulators, hardware accelerators, and simulators which will minimize the verification runtime within the budget.

Given a particular design and set of tests, one can also compute the most cost effective approach by computing the lowest cost per verification frequency. The most cost effective approach for achieving a certain number of verification cycles per second could then be chosen.

A more general notion of cost function can be supported by using the verification runtime as the domain for some function. If the function is monotonically increasing, then one simply needs to minimize the runtime as above.   When the function has inflection points, then the analysis may require linear equation solvers or differential equations. Either way, the performance functions listed above are simple enough to facilitate easy manipulation and optimization.

One can compute the expected performance for a verification technology and determine how long a set of regression tests would take to complete. In contrast, determining the support, maintenance, and depreciation costs for a specific verification technology is subject to many variations with respect to cost accounting.

When considering potential cost functions, one may wish to divide the design and verification process into phases, based largely on the robustness of the components and integrated system. In such a case, one may divide the verification process into an initial development/debug phase, followed by a block   regression   testing   phase,   which   leads   into   a   system development/debug/integration phase, and ending in a system regression phase. The types and number of tests applied, the expected number of errors, and verification budget availability may change for each phase. As the design process continues, the number of errors should decrease while the number of tests and their complexity should increase. Thus, the cost and performance modeling effort may be sub-divided into some number of phases with cost optimization within each phase.

1.4   Results

Having developed performance and cost models, we now use these models to explore the tradeoff space for verification technologies.

To use these models, the designer must supply such parameters as the time required to compile or synthesize a design and to simulate an event or gate. In the case of the Auriga® parallel HDL simulator produced by FTL Systems, the embedded scheduling achieves performance similar to cycle simulation for synchronous systems, requiring approximately five machine instructions per gate simulated, while with discrete event simulation Auriga® requires approximately thirty machine instructions per event (with minor variances for each specific instruction set architecture) [10]. The compilation of a 10 million gate design typically takes five to ten minutes on a 4-way SMP workstation, with incremental compilation (recompiles) usually taking less than a minute. The reader is encouraged to obtain the parameter values for the verification alternatives of interest and perform the same calculations.

The time required to complete a set of verification regression tests is shown in Figure 1, with the following assumptions. First, the design contains 10 million gates and we assume the emulator will compile/recompile the design in five hours and executes at 100KHz. These parameters coincide with commercially available emulators with a cost of approximately $11M. For the hardware accelerator, we assume a compile time of 200 minutes, and that it executes at 1KHz. We assume a cost for the hardware accelerator of $650K. For the cycle simulator, we assume it is executing on an 8-way configuration of 1GHz processors, each of which issues 2.5 instructions per cycle. Assuming 5 machine instructions per gate simulated, the execution frequency is 400Hz. The cost for this simulator and workstation is approximately $100K. For each of the systems, we assume there are 25 bugs to be detected, and that each takes one hour to find and fix. With these assumptions, one can see in Figure 1, that as the total number of simulation cycles increases from one million to one trillion, the emulation and hardware acceleration technologies perform better. For smaller verification suites, the cycle simulator performs best until the number of test vectors approaches one billion.

To take into consideration the large difference in the cost of the verification technologies, Figure 2 shows the performance achieved with a $10M for the hardware acceleration and simulation approaches. These results assume that the regression tests can be divided into subsets and run in parallel. Once again, cycle simulation is the best approach for smaller test sets, with emulation the best approach for test sets exceeding approximately fifty billion vectors.

The number of errors detected in the design impacts the proper choice verification technology. Simulation is best for shorter regression tests involving debugging, while hardware acceleration and emulation work best as the length of tests increases and the need for debugging decreases. If the verification tasks correspond to four phases (initial development/ debug, block regression testing, system development/ debug/integration, and system level regression), then we may expect the number of errors found to decrease as testing progresses. Figure 3 illustrates the different performance levels for cycle simulation, emulation, and hardware acceleration with different numbers of detected errors. As expected, the emulation and hardware acceleration performs best with fewer errors and the cycle simulation performs best when there are more errors. Based on an analysis like this, a designer can determine when to use each of these verification technologies to best meet the schedule and cost needs.

1.5   Conclusions

The importance of verification continues to grow with the size of designs. In this paper we focused on developing and using a performance/cost model to perform tradeoffs between the different types of emulation, hardware acceleration, and simulation tools available to designers. As a tool to help in this evaluation, some simple performance models were developed for each of these verification engines, followed by some possible cost functions to help determine which verification approach best suits the needs of a design group at a particular point in the design process. While exploring some of the results from applying these models, we investigated how each verification technique may be best suited to particular phases of design or cost/performance regimes. By applying models like these, engineers and engineering managers can make quantitative tradeoffs in determining the verification strategy to employ for a given design.

Although some typical values are listed for the parameters used in the models in this papers, readers are encouraged to insert parameters based on their design requirements and the specific hardware and software tools potentially available to them for verification. Further research remains to be completed with respect to refining these models to support more detailed probability distributions used for errors, additional exploration into "typical" parameter values, other possible cost functions, and providing better modeling support for the costs and performance of hardware/software coverification techniques.

1.6 References

[1] Vishwani D. Agrawal and Srimat T. Chakradhar, "Performance Analysis of Synchronized Iterative Algorithms on Multiprocessor Systems." *IEEE Transactions on Parallel and Distributed Systems*, 3(6): 739-746, November 1992.

[2] Richard M. Fujimoto. "Parallel Discrete Event Simulation," *Communications of the ACM*, 33(10):30-53, October 1990.

[3] Michael Keating and Pierre Bricaud, *Reuse Methodology Manual*. Kluwer Academic Publishers, 1998.

[4] Bradley L. Noble, Gregory D. Peterson, and Roger D. Chamberlain, "Performance of Synchronous Parallel Discrete-Event Simulation." *28th Hawaii International Conference on System Sciences*. Waileau, Maui, HI, January 1995.

[5] Athanasios Papoulis. *Probability, Random Variables, and Stochastic Processes*. McGraw Hill, 1984.

[6] Gregory D. Peterson and John C. Willis, "High Performance Hardware Description Language Simulation: Modeling Issues and Recommended Practices." *Transactions of the Society for Computer Simulation*, 16(1):6-15, March 1999.

[7] Gregory D. Peterson and Roger D. Chamberlain, "Parallel Application Performance in a Shared Resource Environment." *IEE Distributed Systems Engineering Journal*, 3(1):9-19, 1996.

[8] Gregory D. Peterson, *Parallel Application Performance on Shared, Heterogeneous Workstations*. Doctoral dissertation, Washington University, St. Louis, MO, December 1994.

[9] Gregory D. Peterson and Roger D. Chamberlain, "Beyond Execution Time: Expanding the Use of Performance Models." *IEEE Parallel & Distributed Technology*, 2(2):37-49, Summer 1994.

[10] John C. Willis and Daniel P. Siewiorek, "Optimizing VHDL Compilation for Parallel Simulation," IEEE Design and Test of Computers, pp 42-53, September 1992.

[11] Kenneth F. Wong and Mark A. Franklin. "Performance Analysis of a Parallel Logic Simulation Machine," *Journal of Parallel and Distributed Computing*, 7:416-440, 1989.

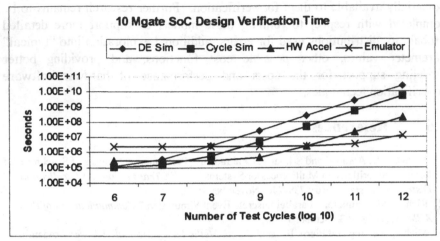

Figure 1. Regression Runtimes for 10MGate Design

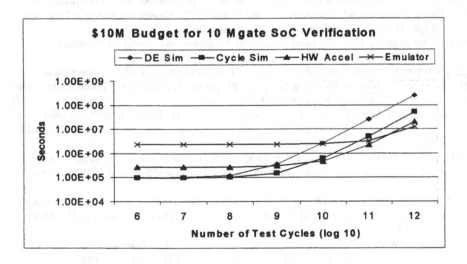

Figure 2. Regression Times with Equivalent Verification Budget

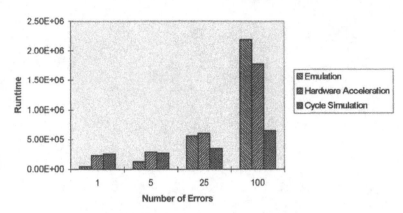

Figure 3. Impact of Errors on Verification Time

TCL_PLI, a Framework for Reusable, Run Time Configurable Test Benches

Stephan Voges and Mark Andrews
(Stephan.Voges@efi.com, Mark.Andrews@efi.com)
Electronics For Imaging, Inc.

Key words: Test bench, Verilog, Tcl, PLI

Abstract: As ASIC complexity keeps increasing, the time spent in design and
maintenance of test benches has grown to become a disproportionately large
part of the total design effort. Verilog test benches have become slow to
compile and cumbersome to maintain.

This paper discusses a scripting approach to managing the test bench
complexity issue. Partitioning the functionality of a test bench between
Verilog and a scripting language allows for a significant reduction in compile
times during ASIC verification. If done correctly, partitioning also offers great
potential for re-use of test bench components.

The Tcl language was chosen as a basis for implementing a library of PLI
routines that allow fully customizable interpreters to be instantiated in Verilog
test benches. This library allows multiple Tcl interpreters to be instantiated in
a Verilog simulation. The Tcl interpreters can interact with the simulation and
cause tasks to be executed in the Verilog simulation.

We have found the TCL_PLI library to be extremely valuable in speeding up
our verification efforts on multi-million gate ASICs and have decided to make
the library available to the general design community. The source code for the
library may be obtained by contacting the authors.

1.   INTRODUCTION

Our traditional approach to test bench development and ASIC
verification was to write a single test bench that contains a number of tasks
that exercise different aspects of the designs. Team members would add
new tasks as the verification effort continued, and tests would be run by

P.J. Ashenden et al. (eds.), System-on-Chip Methodologies & Design Languages, 269–281.
© 2001 *Kluwer Academic Publishers. Printed in the Netherlands.*

calling different sets of tasks in different order, depending on the functionality being tested.

The problem with this approach was that the test bench had to be recompiled for every new simulation. Even though compiled simulators offer features like incremental compilation, in which only the code that changed was recompiled, this still meant that a lot of time was spent compiling test benches.

Our initial efforts to reduce this problem involved using "configuration files." In this approach, we would try to make the test bench code to be "all things to all people." Test benches typically contained complicated case statements and if-then-else sequences. These were controlled by parameters that were read from a text configuration file when the simulation was launched.

This approach eliminated the need to repeatedly recompile test benches, but introduced an even worse problem: test benches became extremely complicated, very difficult to maintain, and very application specific. Very often when new functionality was added to a test bench, it had side effects that caused other tests to break. At the end of the project we had a test bench that very few people understood, and was so convoluted that there was no possibility of reuse.

2. SCRIPTING AS A SOLUTION

We decided that the solution to our dilemma was to make the "configuration files" more powerful, so that some of the complexity that was coded in Verilog could be shifted over to something that was evaluated at run time. In essence, we needed a scripting language that could interact with the simulation.

With this approach, we could design a simple test bench, which would contain a number of tasks that handle low level interaction with the device under test. If these tasks could then be called from a scripting language, we could develop different verification scripts; each tailored to verifying a specific aspect of the design.

The key difference from the configuration file approach is that the intelligence that determines the ordering of events is moved from Verilog to a script file that is interpreted at run time. Different people could then use the same test bench to test a design in completely differing ways. They would have the benefit of significantly reduced need for recompilation, and different people on the verification team would be completely insulated from one another. Each would be developing their own verification script, and changes to that script would affect only their own simulations.

3. TCL AS A CHOICE OF SCRIPTING LANGUAGE

We started out with the idea to extend the configuration files that we were already using to handle simple conditional and looping constructs, and to extend the PLI routines that we were using to process these configuration files to become a simple interpreter.

Fortunately, very early in the project, we decided to investigate the feasibility of using Tcl as our scripting language. This turned out to be a very good idea. It was immediately obvious that John Ousterhout, the inventor of Tcl, had been in our situation many times. It was after many attempts to write custom interpreters for various tools, that he decided to write the ultimate, reusable and embeddable interpreter[1].

Tcl had all the characteristics that we were looking for: It was already developed, it was free, it was easily extendable, and it was easy to embed in an application. In a nutshell: it was invented for this specific purpose.

4. THE OBSTACLE

We only had one problem to overcome: The Verilog language is extended through function calls: If a user needs new functionality, he implements it in C, and uses the PLI API to make it available as a new function that can be called from Verilog. Tcl is also extended through function calls: new functionality is implemented in another compiled language (typically C) and is linked into Tcl as a new Tcl function.

We wanted to implement a system in which the Verilog simulation could start up a Tcl interpreter and instruct it to run a script. This implied a PLI function call. At some point the Tcl interpreter would encounter a function that is mapped to a certain Verilog task. It then has to pass control back to Verilog so that the task could be executed. This implies that the PLI call that invoked the interpreter has to return, leaving the problem of how to resume execution of the Tcl script after executing the Verilog task.

We had to figure out a system through which we could pass control between Verilog and Tcl, through function calls on either side, so that the Verilog code controls when Tcl interpreters are invoked, but that the Tcl interpreters could cause Verilog tasks to be executed (implying that a PLI call needs to return), while retaining the state of the Tcl interpreter so that it could resume execution of the Tcl script after the Verilog task completed. Figure 1 illustrates the required interaction between the Verilog simulation and a Tcl interpreter.

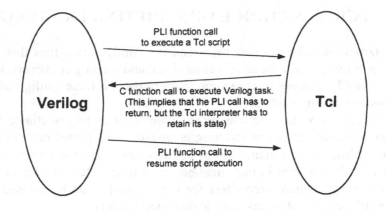

Figure 1. Required interaction between Verilog and Tcl

5. THE SOLUTION

Our solution to this problem was to implement a simple client-server model, in which the Tcl server runs on a separate thread from the Verilog simulation. Synchronization between the Verilog simulation and the Tcl server is done via a set of semaphores. This allows control to be passed freely between Verilog and Tcl. The Verilog code determines when Tcl interpreters are invoked and Tcl interpreters can randomly cause Verilog tasks to be executed.

The TCL_PLI library allows any number of Tcl interpreters to be instantiated in a Verilog simulation. Every interpreter is completely customizable. PLI functions are used to initialize interpreters and define new Tcl functions that are mapped to Verilog tasks. PLI functions are also used to start running scripts on the Tcl interpreters and to pass control between Verilog and Tcl.

The Verilog tasks have access to arguments passed to the Tcl functions that invoked them, allowing information to be passed from Tcl to Verilog. The Verilog tasks also have the ability to control the return values of their Tcl counterparts, allowing information to be passed from Verilog to Tcl. PLI routines are also provided that allow direct sharing of information between Tcl and Verilog, as well as between different Tcl interpreters.

6.   INTERACTION BETWEEN VERILOG AND TCL

At the core of the TCL_PLI library are five PLI functions: $tclInit, $tclExec, $tclGetArgs and $tclClose.

$tclInit is used to create and initialize a new Tcl interpreter. It defines the new Tcl functions that will be used to invoke Verilog tasks, maps them to specific tasks, and defines how many arguments they will take.

$tclExec is used to pass control from Verilog to the Tcl server. It is used to launch a new script or to resume execution of a script that was stalled when the interpreter encountered a function that was mapped to a Verilog task. $tclExec will return under one of three conditions: when an error occurs, when the script ends or when a function is encountered that is mapped to a Verilog task.

$tclGetArgs is used to access the argument values that were passed to an extended Tcl function. $tclClose is used to destroy a Tcl interpreter and free the resources associated with it.

The following code snippet shows how an interpreter is created and initialized in Verilog. The interpreter will have three extended Tcl commands (b_write, b_read and b_wait_irq) that are mapped to Verilog tasks. These commands will allow scripts running on the interpreter to write data on a bus, read data from a bus or wait for an interrupt to occur on the bus.

```
 1 parameter BUS_WRITE      = 1,
 2             BUS_READ      = 2,
 3             BUS_WAIT_IRQ  = 3;
 4 initial
 5    begin: processor_model
 6    integer tcl_handle, tcl_command,
 7       tcl_return_value;
 8    // Initialize interpreter that knows about
 9    // the three extended Tcl commands:
10    tcl_handle = $tclInit ("processor",
11       tcl_command, tcl_return_value,
12       "b_write",   BUS_WRITE,   2,
13       "b_read",    BUS_READ,    1,
14       "b_wait_irq", BUS_WAIT_IRQ, 0
15       );
16    if (tcl_handle == 0)
17       begin
18       $display ("Init error.");
```

```
19          $finish;
20          end
```

On lines 1 to 3, three integer parameters are defined that represent the three extended Tcl commands in Verilog. Lines 6 and 7 define three variables that are used to communicate between Verilog and Tcl. tcl_handle will be used to store the handle of this interpreter. Each interpreter will have a unique handle. It is returned by the $tclInit PLI function and is used by all other TCL_PLI functions to identify the interpreter that is being addressed. tcl_command will be used by TCL_PLI to indicate to Verilog which extended Tcl function had been encountered while executing the script. It is also used to indicate the completion status of the Tcl script when execution of the script ends. tcl_return_value will be used by Verilog to communicate the return values of tasks back to the calling Tcl functions.

On lines 10 to 15, the call to $tclInit initializes the interpreter. It identifies the variables tcl_command and tcl_return_value to TCL_PLI. It also defines the extended Tcl functions b_write, b_read and b_wait_irq. $tclInit associates an integer value with each extended Tcl function and stores the number of arguments that each extended Tcl function should expect. $tclInit can take a variable number of arguments to allow definition of any number of extended Tcl functions.

The return value of $tclInit will contain the handle of the newly created Tcl interpreter. If the value is zero, this is an indication that an error occurred while creating and initializing the interpreter. The test on line 16 confirms that the call to $tclInit completed successfully.

The next code snippet indicates how a script is executed on a Tcl interpreter and how control is passed between Verilog and Tcl.

```
21      while ($tclExec (tcl_handle,
22          "example.tcl"))
23        begin
24        tcl_return_value = 0;
25        case (tcl_command)
26          BUS_WRITE:
27              bus_write (tcl_handle);
28          BUS_READ:
29              bus_read (tcl_handle,
30                  tcl_return_value);
31          BUS_WAIT_IRQ:
32              bus_wait_irq;
33        endcase
34        end // while
```

```
35    if (tcl_command != 0)
36        begin
37            $display ("Error in Tcl script!");
38            $finish;
39        end
40    $tclClose (tcl_handle);
41    end // processor_model
```

When $tclExec is encountered the first time (line 21), the Tcl server is idle. This is an indication that it should start executing the script "example.tcl". As mentioned earlier, $tclExec will return under one of three conditions: if an error occurs, if the script ends or if an extended Tcl function is encountered. A non-zero return value indicates that an extended Tcl function had been encountered. A return value of 0 indicates that the script has ended or that an error occurred. In the case where an extended Tcl function was encountered, the Tcl interpreter will stall until the next call to $tclExec informs it that the Verilog task has been executed.

The while loop (lines 21 to 34) will keep looping as long as the return value of $tclExec remains positive. This means that the simulator will keep executing Verilog tasks as they are encountered in the Tcl script that is being executed. If $tclExec is called when the Tcl server is not idle, it assumes that execution of the current script should continue. $tclExec will check that the script name passed to it matches the name of the script being executed and will return an error value if this is not the case.

Not all Verilog tasks have meaningful return values. Therefore tcl_return_value is initialized to zero to ensure that an undefined value is not eventually returned to the Tcl interpreter (line 24).

Whenever $tclExec returns a non-zero value, the tcl_command variable will contain the integer value corresponding to the extended Tcl function that was encountered in the script. The case statement on line 25 is used to call the Verilog task associated with this Tcl function.

Once the Verilog task completes, the while loop continues with another call to $tclExec. $tclExec will again return when it encounters an extended Tcl command, reaches the end of the script or encounters an error. The result is that the while loop will cause the whole Tcl script to be executed, and Verilog tasks will be called in the order determined by the execution of the Tcl script.

When the while loop terminates, it is appropriate to check that no error occurred (line 35). When $tclExec returns zero (causing the while loop to terminate), the tcl_command variable will contain an exit code indicating either normal termination (end of script reached) or abnormal termination (due to an error in the script).

At this point, the Tcl interpreter can be deleted if it is no longer required (line 40). It should be noted that the same interpreter can be used repeatedly. Scripts can also be run on the interpreter from different points in the Verilog code. TCL_PLI will generate an error if an attempt is made to run multiple scripts on one interpreter simultaneously. Interpreter state information (like variable values and procedure definitions) is preserved until the interpreter is destroyed through a call to $tclClose.

The next code snippet shows the definition of the Verilog tasks that are called under control of the Tcl interpreter.

```
42 task bus_write;
43     input [31:0] tcl_handle;
44     integer address, data;
45       begin
46         $tclGetArgs (tcl_handle,
47           address, "i",
48           data,    "i"
49         );
50     // Simulate the write cycle on the bus
51     end
52
53 task bus_read;
54 input [31:0] tcl_handle;
55 output [31:0] data;
56 integer address;
57     begin
58     $tclGetArgs (tcl_handle,
59       address, "i"
60       );
61     // Simulate the read cycle on the bus
62     data = data_read_from_bus;
63     end
64
65 task bus_wait_irq;
66     begin
67     @(bus_irq);
68     end
```

A Verilog task can gain access to the arguments of the Tcl function that invoked it. For this purpose the handle of the interpreter is passed to these tasks (line 43). A call to $tclGetArgs is used to transfer this information from Tcl to Verilog (line 46). $tclGetArgs can handle integer or string arguments, and will do the appropriate conversions.

The bus_read Verilog task illustrates how the return value of a Tcl function is set up in the Verilog task (lines 62 and 30). TCL_PLI assumes that the return value is an integer. The bus_wait_irq task illustrates a simple case where the Tcl interpreter can be stalled while waiting for an event in the simulation (line 65).

Following is a sample Tcl script that can be run in the Verilog example shown above. It illustrates how the execution order of the Verilog tasks is completely controlled from Tcl. In essence, the Tcl script is in complete control of the simulation in the same way that software controls the hardware on which it is run.

```
1  # Write to a register, wait for an interrupt
2  # and read back a cause:
3  puts "$::vname @ $::vtime: \
4      Writing 0xaa to address 0x05:"
5  b_write 0x05 0xaa
6  puts "$::vname @ $::vtime: \
7      Waiting for interrupt..."
8  b_wait_irq
9  puts "$::vname @ $::vtime: \
10      Interrupt received"
11 puts [format "Address 0x05 now contains %x" \
12      [b_read 0x05]]
13
14 # Write to another register, then poll until
15 # bit 1 changes:
16 puts "$::vname @ $::vtime: Writing 0xff to \
17      address 0x0a:"
18 b_write 0x0a 0xff
19 set bit_1 0x1
20 while {$bit_1} {
21     # Read 0x0a and check the LSB:
22     set bit_1 [expr [b_read 0x0a] && 0x01]
23     puts "$::vname @ $::vtime: Bit 1 value \
24         is $bit_1"
25 }
26 puts "$::vname @ $::vtime: Value changed!"
```

The variables vname and vtime are defined by the TCL_PLI library. vname contains the name of the interpreter (passed as the very first argument to $tclInit). It is useful to determine the source of a message.

vtime contains the current simulation time, which is very useful for tracing simulator messages back into waveforms.

7.   BEHIND THE SCENES

As mentioned previously, the Tcl interpreter is run on a separate thread from the Verilog simulation. A call to $tclInit will cause a secondary thread to be created, on which the Tcl server will run. The Tcl server creates and initializes a new Tcl interpreter, and then enters a loop in which it waits for and executes commands received from the PLI functions.

The following C code snippet is a simplified version of the command loop in the Tcl server.

```
1  /* Wait for and service requests to run
2     scripts */
3  runServer = 1;
4  while (runServer)
5    {
6    /* Pass control to the Verilog thread */
7    sem_post (&t->t2v);
8    /* Wait for an instruction from the
9       Verilog thread */
10   sem_wait (&t->v2t);
11   switch (t->serverCommand)
12     {
13     case TC_RUNSCRIPT:
14        t->serverStatus =
15           tclServer_runScript (t);
16        break;
17     case TC_CLOSE:
18        runServer = 0;
19        t->serverStatus = TS_DONE;
20        break;
21     case TC_ADDCOMMAND:
22        /* Add new commands to
23           interpreter */
24        break;
25     case TC_LINKVARS:
26        /* Link with Verilog variables */
27        break;
28     case TC_SHAREVARS:
29        /* Link with variables from other
```

```
30                      interp */
31           }
32     }
```

Once the Tcl server has initialized, it enters the command loop and immediately posts the t2v semaphore to the PLI to indicate that it is ready to accept a command. It then waits for the v2t semaphore. Upon receipt of the v2t semaphore, it examines the serverCommand member of its defining structure to determine what command was issued and executes the command. When the command is completed, the loop starts again.

On the Verilog thread, a PLI function sends a command to the Tcl server by setting up the serverCommand member of the server's defining structure and then posts the v2t semaphore. The PLI function then waits for the t2v semaphore as an indicator that control is being passed back to Verilog.

The following sequence will take place if the Tcl script in the previous example is executed: When $tclExec is called from the Verilog simulation the first time (Tcl server is idle), it will instruct the Tcl server to start executing the script, and then wait for the t2v semaphore. The Tcl server will execute lines 1 through 4 of the script. When it reaches line 5, which contains the extended command b_write, it will call the function tclServer_verilogCall. This function is called for all extended commands that are mapped to Verilog tasks. tclServer_verilogCall will save the relevant information in its defining structure, post the t2v semaphore and then wait for the v2t semaphore.

When $tclExec receives the t2v semaphore, it will synchronize linked variables and return to Verilog. This allows the simulator to enter the while loop in which it executes the tasks associated with extended Tcl functions (lines 21 to 34 of the Verilog example). When the task associated with b_write completes, $tclExec is again called. This time the Tcl server is not idle, so $tclExec assumes that execution of the script should resume. It will synchronize linked variables, post the v2t semaphore and wait for the t2v semaphore.

tclServer_verilogCall will receive the v2t semaphore and return, allowing execution of the Tcl script to continue. This process will repeat itself until the script completes. When this happens, the Tcl server will indicate this to the PLI when posting the t2v semaphore. This time, when $tclExec returns, the Verilog while loop will terminate.

The following C code snippet shows a simplified version of tclServer_verilogCall, the function that is called by the Tcl interpreter when it encounters an extended command.

```
1  int tclServer_verilogCall (...)
2  {
3      /* Store the command value for the Verilog
4         thread */
5      t->commandValue = command->value;
6      /* Store the argument array and count in
7         the interpreter struct to make it
8         accessible to tclGetArgs: */
9      t->argc = argc;
10     t->argv = argv;
11     /* Semaphore verilog */
12     sem_post (&t->t2v);
13     /* At this point control has been passed
14        back to Verilog, where the
15        functionality is being simulated.  Once
16        done, the main thread will semaphore
17        this thread to continue. */
18     /* Wait for semaphore from verilog */
19     sem_wait (&t->v2t);
20     /* Set up the return value */
21     sprintf (message, "%d", t->retVal);
22     Tcl_SetResult (t->interp, message,
23        TCL_VOLATILE);
24     return TCL_OK;
25  }
```

It is important to note that, even though TCL_PLI is multi-threaded, and that every interpreter is run on a dedicated thread, the essential single threaded nature of Verilog simulations is maintained. Only one call to $tclExec can be reached in the Verilog simulation at any given time. The Verilog simulation stalls until this call returns, which will happen when the Tcl interpreter calls tclServer_verilogCall or when the script completes. tclServer_verilogCall, on its part, will only return when the next call to $tclExec is encountered in the Verilog simulation. This means that, even though many Tcl scripts may be in the process of execution at any given moment in time, only one of them or the Verilog code itself will be running at that moment. All event scheduling and execution order is still under the control of the simulator.

It should also be noted that the Tcl server executes scripts in zero simulation time. Simulation time doesn't advance for the duration of a call to $tclExec. Simulation time advances normally while the Verilog tasks invoked from Tcl are executed

8.   CONCLUSION

The solution that we devised more than fulfilled our expectations: We managed to significantly reduce time spent recompiling test benches. Our test benches are a lot simpler, and consist of modular, reusable modules that are easy to maintain. New tests are implemented in new, separate scripts, eliminating the problem where addition of new tests caused existing tests to break.

One of the greatest advantages was not anticipated: The Tcl scripts themselves are often reusable. When we do module level testing, we partition functionality between Tcl and Verilog so that the Tcl scripts will be reusable in system level testing. With careful partitioning, it is only necessary to rewrite the Verilog tasks that implement the extended Tcl functions to interact correctly with the system level interface. This way the scripts developed for testing the module can be rerun without modification in system level tests.

In one case, we were able to reuse a complicated Tcl script written for a project that was cancelled! It was initially written to test a memory controller by doing random reads and writes and automatically verifying data integrity. When we later had to design logic that had to access system memory through a PCI bus, we reused the script without modification.

Another unexpected benefit was that Tcl suddenly became a mainstream language in the EDA world. When Synopsys decided that they were going to integrate Tcl interpreters into their synthesis and static timing analysis tools, we could immediately take advantage of this, because our whole design team was already familiar with Tcl.

We are currently investigating the feasibility of porting our Tcl scripts to real hardware. This will enable us to run our verification suite on our ASICs when they return from the foundry. For PCI based devices, this seems like it will be a simple task. We just need to write simple C functions that map our PCI read and write commands to real PCI read and write commands.

All in all, our decision to switch to scripting languages to manage test bench complexity proved to be a very good one. In our opinion, choosing Tcl as the scripting language meant the difference between success and failure.

REFERENCES

[1] John Ousterhout, "Tcl and the Tk Toolkit," p. xvii, Addison-Wesley, 1994.

METHODS FOR SOC DESIGN AND RE-USE

Object-Oriented Specification and Design of Embedded Hard Real-Time Systems

Wolfgang Nebel, Frank Oppenheimer, Guido Schumacher,
Laïla Kabous, Martin Radetzki, Wolfram Putzke-Röming
Carl von Ossietzky Universität Oldenburg and OFFIS,
PO box 25 03, 26111 Oldenburg, Germany
Phone: (+49) 441-798-4519, Fax: (+49) 441-798-2145
email: nebel@informatik.uni-oldenburg.de

Key words:

Abstract:

The approach presented in this paper is a contribution to combine well established
methodologies in software engineering, namely object-orientation, with novel co-simulation
techniques for real-time HW/SW system simulation, and new synthesis techniques based on
the HW-semantics of object-oriented HW-specifications. The result is a consistent object-
oriented design process for safety critical hard real-time embedded systems. Its objective is to
decrease the design time of such systems by improving the re-usability, enabling concurrent
HW/SW co-design and avoiding unnecessary iteration loops to meet timing constraints. The
paper presents a motivation for and a survey of the methodology as well as a brief
introduction into the different phases of the process, in particular simulation and synthesis.
The concepts are illustrated by a consistent example taken from an established system level
benchmark.

1. INTRODUCTION

The penetration of microelectronics and information processing
technology in technical products is obvious to everyone today. An example
is the automotive industry with an electronic value content between 20%
(compactcars) to 40% (luxury cars). Such embedded systems consist of

285

P.J. Ashenden et al. (eds.), System-on-Chip Methodologies & Design Languages, 285–296.
© 2001 *Kluwer Academic Publishers. Printed in the Netherlands.*

dedicated HW components and application software which need to be developed concurrently in order to meet time to market requirements.

Most embedded systems, e.g. in automotive, avionics, and medical applications are extremely safety critical with respect to functional and non-functional requirements, in particular hard real-time (HRT) constraints. These safety requirements demand a design process which ensures to the largest possible extent a provable correct implementation. Thirdly, the complexity of both HW and SW ask for a reuse based design flow.

Putting these requirements on a design methodology together, it is almost self-evident to consider object-oriented methods for a HW/SW co-design process. This can improve the re-usability through its encapsulation mechanisms and avoids the paradigmatic break between OO-SW-engineering and block oriented HW-design. For HRT systems, however, a construction methodology is needed in addition, which supports a schedulability analysis in co-design. Such a methodology is in use for SW-design in the avionics industry today, namely HRT-HOOD.

In this paper we present a generic consistent object-oriented co-design flow for embedded HRT systems. It is based on the principles of the HRT-HOOD methodology which has been extended to co-design by a co-simulation framework and by OOHARTS, a real-time extension of UML. In order to provide an implementation interface to dedicated HW, synthesis concepts are presented how to translate object-oriented HW-specifications into HW-models described in standard VHDL.

The design flow is partly illustrated with a benchmark example presented in Section 2 of the paper. The principle requirements on HRT-System design are discussed in Section 3 together with a survey of existing approaches and a brief introduction to OOHARTS and HRT-HOOD. Section 4 then introduces the main phases of the proposed design flow, before presenting the co-simulation framework in Section 5. The HW semantics of object-oriented descriptions and the synthesis approach are described in Section 6. We conclude in Section 7 with a brief survey of current and an outlook on intended projects supplementing and completing the proposed design methodology.

2.   BENCHMARK

We will use the portal crane controller case study [9] to illustrate some of the methods proposed in this paper. This benchmark has been introduced for a panel discussion about modelling languages at DATE'99 and at FDL'99.

A control system drives a crane car along a track to carry a load to an arbitrary position within the plant depicted in Figure 1. Since the cable between the car and the load is flexible, the load will start to swing when the car accelerates. This behaviour is specified through a system of differential equations describing the analog physical behaviour of the cranes car and its load.

The algorithms to calculate the correct values for the actuators are described very detailed in the specification [9]. The control algorithm tries to dampen the swinging of the load when the desired position is nearly reached. Other control activities have a rather simple functionality but their real-time requirements are very strict.

In order to validate the correctness and robustness of the model three test cases are provided with the benchmark. Each of them proves different properties of the modelled crane controller.

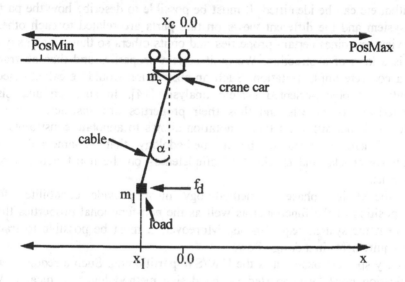

Figure 1: Schematic picture of the portal crane

The benchmark tries to combine typical characteristics of embedded system design like mixed analog/digital behaviour, HW/SW co-design, concurrence of control tasks and hard real-time requirements.

3.   HRT-SYSTEM DESIGN

The design of hard real-time systems like the one presented in the previous section introduces requirements on the design process that differ from traditional software engineering. The differences are mainly imposed by non-functional requirements like e.g. timing constraints that characterise hard real-time systems.   Dealing with such non-functional requirements is a crucial issue in the specification and design of a hard real-time system.

In this section, we discuss the requirements on design methodologies for hard real-time systems.

3.1   Requirements

If we want to specify and design an embedded system, the first design step is to derive a formal description which defines the requirements of the system from an informal description. To describe the requirements, a notation has to be used that must be able to express all the properties of the system to be designed.

In an initial description, different parts, entities, or components of a system can be classified. In the crane example, sensors, actuators, a control algorithm etc can be identified. It must be possible to describe how the parts of a system and the different views on that parts are related to each other. Each view examines certain properties and omits others so that we can say a view is a kind of abstraction. We omit certain properties and thus abstract from a concrete implementation. Such an abstraction would be called object in traditional object-oriented system analysis [14]. In such an analysis, characteristics of objects and thus their properties are abstracted by attributes. It is not sufficient that a notation allows to annotate constraints to objects. There must rather be a methodology that is amendable to consistency checks and checks for completeness on the non-functional requirements.

In the design phase a methodology must provide capabilites for decomposition of the functional as well as the non-functional properties that implement the system requirements. Moreover, it must be possible to trace these requirements in a design flow.

A very special constraint is the HW/SW partitioning. Such a requirement specification must be supported in the design methodology[1]. Typical HW/SW communication and synchronisation schemes via e.g. device registers and interrupts, should be available in the design methodology.

The implementation step that consists of coding and hardware synthesis imposes platform-specific or technology specific environment constraints. The design methodology must support a backannotation step that allows to check the new constraints against the ones that already exist.

So far, we have presented a list of requirements on design methodologies for hard real-time systems. In the next section we discuss to what extent current design methodologies meet these requirements.

---

[1]   We address a system complexity here that is not supported by automated HW/SW partitioning concepts.

3.2   State of the art

In recent years object-oriented techniques have been employed in the design and development of real-time systems. Examples of OO methods which are used for the construction of real-time systems are ROOM [13], UML-RT [12], OCTOPUS [1], Real-Time UML [5], etc.

The methodologies mentioned above have been developed with various kinds of real-time systems in mind and have particular strengths and weaknesses. But in particular for the design of hard real-time systems, they do not fully support the explicit specification of the timing requirements and constraints. They lack a validation of the system models designed in early design phases and fail to generate analyzable models in order to check if the timing requirements of the specification are met by the design.

In the following we present two design methodologies which are particularly developed for the design of hard real-time systems that emphasize the provability of real-time behaviour.

3.2.1   OOHARTS approach

The OOHARTS approach [8] is based on the Unified Modeling Language UML [11], which is established as a standard for object oriented analysis providing notations for the specification, visualisation, construction and documentation of software systems, and the hard real-time concepts of HRT-HOOD [4].

The development process of OOHARTS is an incremental and iterative one. It consists of four main phases. The first phase covers the definition of the system requirements. During this design step, the functional and the non-functional properties extracted from the problem description of the system are addressed. The second phase defines the structure and behaviour of the system. We use the UML class diagrams for the definition of the system structure. We have introduced new notations with a formally defined semantics into UML to describe the objects with their real-time (non-functional) requirements common to hard real-time systems. The definition of the objects' behaviour uses an extended form of the UML state diagrams [11] to explicitly describe the behaviour of the objects including their timing constraints.

The third phase is the design phase. This phase is the process of specifying an implementation that is consistent with the results of the analysis. It consists of a refinement of the objects and models defined during the analysis and of an explicit specification of the communication, synchronisation, and data exchange between the objects. The last phase in the development process of OOHARTS is the implementation.

3.2.2   HRT-HOOD

HRT-HOOD [4] is a design methodology for hard real-time systems that was a developed in a research project commissioned by the European Space Agency and has proven its applicability to elaborate systems in case studies, e.g. in [3]. In this section we present the basic ideas of this design methodology.

A key feature of HRT-HOOD is that it allows to capture non-functional requirements that were characteristic of hard real-time. The design methodology provides appropriate abstraction for these requirements. The requirements are specified as object-attributes.

The explicit abstraction of the non-functional requirements as real-time attributes makes the system specification amenable to scheduling analysis.

Accordingly, the hierarchical decomposition of the system aims at keeping real-time attributes consistent. The idea is to design the decomposed system *correct by construction*.

In HRT-HOOD the implementation language is Ada [7]. With its strong encapsulation and abstraction concepts it directly supports programming in the large as it is required for complex systems. All constructs required for programming concurrent real-time systems, like tasking, communication, access to a clock, delay mechanisms, timeouts, asynchronous transfer of control, interrupt mechanisms, and portable low-level programming concepts are part of the language.

3.2.3   RTOS-based implementation

If a language like C that has no inherent real-time facilities is used as an implementation language, a real-time operating system (RTOS) would be necessary to provide primitives for real-time programming. In this case the abstraction of the execution environment constraints must be deduced from the documentation of the RTOS and the construction of the run-time library. Although this way appears to be industrial practice in several application domains it should be noted that using an approach that is based on an implementation language which includes concurrency leads to more readable, portable and maintainable programs [2].

3.3   Conclusion

Classic object-oriented design methodologies do not meet the requirements for hard real-time systems. It is not sufficient for a design methodology just to capture functional requirements at system level and to design a functional model as system level specification. Managing non-functional system requirements in a design process including system decomposition and refinement is essential in an appropriate design methodology.

Using an implementation language like Ada in an object-oriented design approach like HRT-HOOD avoids extra design steps since the concurrency notions and more generally the real-time notions of HRT-HOOD are at a higher semantic level than most operating system primitives [6].

4. GENERIC DESIGN FLOW

We now present a generic design that improves current software-centric development approaches. The initial system level specification is an object-oriented model with objects that explicitly specify the real-time requirements as attributes. In the following design steps, the system undergoes a hierarchical decomposition. Execution environment constraints are introduced as new real-time attributes. The decomposition rules follow the principle *correct by construction* with respect to functionality and real-time constraints. The software objects are translated into a concurrent program and the hardware objects are translated into synthesisable hardware. The resulting model, consisting of hardware and software objects, can be viewed as an executable specification that can be simulated in a co-simulation that simulates the functional and the non-functional (real-time) properties of the system. It is important to note that, although real-time properties are simulated, the simulation is done before any synthesis steps. Likewise the software program is simulated without any cross-compilation. In the initial design cycle some of the real-time properties are estimated for this early simulation; in further design cycles backannotated values can be used.

The simulation must not be mixed up with a simulation after synthesis and cross-compilation using cell library models and target processor models or target processor simulators.

An improved design methodology includes hardware design and HW/SW interface construction. Including hardware design means to make device modelling part of the design methodology. The design of corresponding device drivers is eased by the HW/ SW interface construction concepts, which considers real-time constraints.

A concrete development approach that realizes the presented generic design flow is described in the next section.

5.   OO-COSIM

The preceding chapters described the characteristics of a design process for embedded hard real-time systems. OO-Cosim[2] provides a set of methods and tools for such a design flow. It enables the designer to capture the hard real-time design, validate the behaviour in an real-time co-simulation model and supports synthesis of hardware and software by a detailed but target independent specification of the embedded system. Through its object-orientation OO-Cosim is capable to handle even complex concurrent real-time embedded systems.

From a rather abstract design the final system can be derived through refinement of its component. These components, we call them design objects, can be traced through the entire flow depicted in Figure 2. Their properties like provided methods, real-time requirements or communication behaviour are propagated from the first specification to the final implementation.

5.1   OO-Cosim Design Flow

The design starts with a HRT-HOOD [4] specification at system level. Following the refinement rules of HRT-HOOD this step ends up in a design which enables schedulability analysis very early in the design flow.

During the next step the system is partitioned, i.e. the design objects are identified as hardware or software components. The software objects are modeled in Ada95 [2] while the hardware is modeled in VHDL. Along with the functional behaviour both languages contain clear semantic concepts to describe concurrency and real-time behaviour. The communication between hardware and software is modeled by device registers and interrupts. The HW/SW co-simulation, a core tool of OO-Cosim, enables the validation of these models in interaction in an executable specification. Both models can be executed in a debugger to enable deep analysis of the combined system.

[2]   This work has been funded by the Deutsche Forschungsgemeinschaft (DFG) under grant number NE 629/4-2.

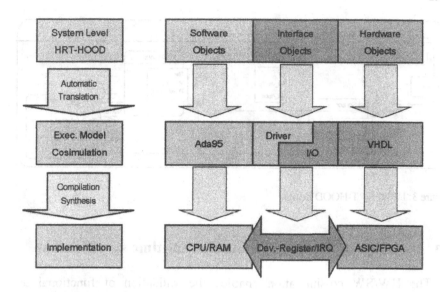

Figure 2: OO-Cosim Design Flow

5.2   Real-time system design

The initial description in HRT-HOOD identifies the design objects, their interfaces, data flow and their relationships. At this early stage the designer specifies the real-time requirement by object-attributes and the communication behaviour of the entire system independently of the later implementation. For the crane benchmark described in Section 2 we identified five main design objects simplified depicted in Figure 3. These objects were refined to more specific objects, e.g. the actuators object was decomposed into objects carrying the methods and real-time attributes of the cranes actuators.

At a rather high abstraction level we identified the objects to be implemented in hardware and software respectively. These are e.g. the „Vc_Actuator" and the „Vc_Regulator" controlling the electrical crane engine and the „Observer" for the safety checks in the hardware.

The „Plant_Environment", the „Operator" and the „Job_Control" were implemented in software because their functionality is rather complex and does not have tight deadlines.

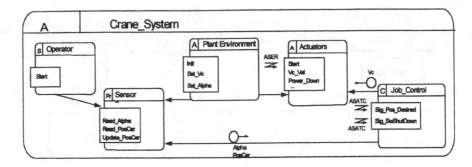

Figure 3: Initial HRT-HOOD design

5.3   Object-oriented executable real-time specification

The HW/SW co-simulation enables the validation of functional and temporal behaviour of the system. At this stage in the design flow the designer wants to observe the detailed behaviour of the specified model. The encapsulation provided by the object-oriented languages increased our models modularity. Thus we could develop hardware and software independently and then integrate it very quickly.

In the crane example we found it extremely useful to check the model behaviour in the co-simulation. We found several modelling errors checking the model against the expected behaviour. E.g. the start up sequence of the embedded system was unspecified and thus error prone.

The communication between software and hardware components was rather complex because numerical values of high accuracy needed to be exchanged via device registers. Only when hardware and software are visible in the simulation, testing can be successful.

Through simulation we also found inconsistencies and mistakes in the benchmark specification, for example, the shutdown button was not active while the crane waited for a new desired position and the crane could not work starting from any other position than xc = 0.0.

6.   SYNTHESIS

To explain hardware synthesis from object-oriented models, we briefly outline the basic concepts:

- Mapping of object data onto a register, register bank, or memory, controlled by user-defined constraints. The register option is preferable if different objects shall have a local state storage of their own so as to be able to operate in parallel. The memory option, on the other hand, is suitable for array data structures, e.g. a vector of the control algorithm whose elements are accessed sequentially.

- Implementation of methods as datapath circuits. These datapaths can read and modify the object's state and have inputs and outputs corresponding to the methods' parameters. If a method's implementation is defined as a complex algorithm, the datapath may have to be sequential. Methods are executed in a mutually exclusive way so as to avoid resource conflicts related to the common use of the state storage. This enables resource sharing among the individual method datapaths.
- Allocating resources for inter-object communication. This includes wires for the transmission of a binary identifier of the method whose execution a client object requests from a server object, of the method's input and output parameters, and of protocol data. Since a server object executes a single method at a time, an arbitration circuit must be inferred when there are multiple clients that can issue requests concurrently.
- Implementing an FSM that controls the method datapaths and executes the communication protocol with clients.

To make object synthesis practicable and provide an interface to existing design flows, we generate VHDL code from the object-oriented model for further processing with a high-level synthesis tool. The object synthesis tool has to generate the required clocking and reset scheme as well as communication and control code in VHDL and to map the abstract data structures of a class to an optimized bit-level representation that facilitates efficient synthesis [10].

7. CONCLUSIONS AND FUTURE WORK

We have presented a consistent fully object-oriented design flow for embedded hard real-time systems. The main contributions are a co-simulation approach for object-oriented HW/SW-systems with real-time constraints, HW-semantics and a synthesis technology for object-oriented HW-specifications and real-time extensions to the UML notation. The system construction follows the decomposition rules of HRT-HOOD, a well established design methodology in safety critical embedded systems.

The current state of the design flow and tool implementation suffers from the missing automated transformation from HRT-HOOD to HW, in particular object-oriented specifications of HW. We will spend efforts to investigate possible formal transformation steps to close this gap. The work on OOHARTS will be extended to investigate a coupling of the HRT-HOOD methodology and UML. In a project funded by the European Commission we are developing a new object-oriented synthesis tool, which will generate output for backend synthesis tools starting from object-oriented models written in SystemC.

8. REFERENCES

[1] Awad, M.; Kuusela, J.; Ziegler, J.: Object-Oriented Technology for Real-Time Systems: A Practical Approach using OMT and Fusion. Prentice Hall, 1996.

[2] Barnes, J.: Programming in Ada95. Addison-Wesley, 1995.

[3] Burns, A.; Wellings, A.; Bailey, C.; Fyfe, E.: The Olympus Attitude and Orbital Control System: A Case Study in Hard Real-time System Design and Implementation. University of York, Department of Computer Science, Internal Report, 1993. Available on the WWW from URL: ftp:// ftp.cs.york.ac.uk/reports/YCS-93-190.ps.Z

[4] Burns, A.; Wellings, A.: HRT-HOOD: A Structured Design Method for Hard Real-Time Ada Systems. Elsevier, 1995.

[5] Douglass, B. P.: Real-Time UML: Developing Efficient Objects for Embedded Systems. Addision Wesley, 1997.

[6] HOOD Technical Group, Rosen, J.-P. (Ed): HOOD An Industrial Approach for Software Design. HOOD Technical Group, 1997.

[7] ISO/IEC 8652:1995(E) Ada Reference Manual, Language and Standard Libraries, Version 6.0, 1994.

[8] Kabous, L.; Nebel, W.: Modeling Hard Real Time Systems using UML: The OOHARTS Approach. UML'99 - The Unified Modeling Language beyond the Standard. Lecture Notes in Computer Science. Springer Verlag, 1999.

[9] Moser, E.; Nebel, W.: Case study: System Model of Crane and Embedded Control. In: Proceedings of the Design Automation and Test in Europe, DATE '99. IEEE Computer Society, 1999, pages 721-723.

[10] Radetzki, M.: Synthesis of Digital Circuits from Object-Oriented Specifications. Dissertation, Oldenburg University, 2000.

[11] Rational Software: Unified Modeling Language UML. Available on the WWW from URL http:// www.rational.com/uml/index.shtml

[12] Rumbaugh, J.; Selic, B.: Using UML for Modeling Complex Real Time Systems. Available on the WWW from URL http:// www.objectime.com

[13] Selic, B.; Gullekson, G.; Ward, P.: Real-Time Object Oriented Modeling. Wiley, 1994.

[14] Shlaer, S.; Mellor, J. S.: Object-Oriented System Analysis Modeling the World in Data. Yourdon Press, 1988.

System Level Design for SOC's
A Progress Report, Two Years On

Grant Martin and Bill Salefski*
Cadence Design Systems, San Jose, California

(* *now with Chameleon Systems, San Jose, California*)

Key words: HW-SW co-design, SOC, system level design, function-architecture co-design

Abstract: In 1998, we discussed design methodology and tool requirements that would allow designers to create IP-dominated System-On-Chip (SOC) devices at the *system* level of abstraction. Here we discuss the lessons learned over the course of the last two years in developing system level design tools, now commercially available, which support the function-architecture co-design approach for complex, platform- and non-platform-based SOCs. In particular, we draw on the experiences of design partners in applying the methods and tools to their applications. Finally, we discuss several lessons learned, which point to future development directions.

1.   INTRODUCTION

In 1998, [1,2] we described the design methodology and tool requirements that would allow designers to create IP-dominated System-On-Chip devices at the *system* level of abstraction – going well beyond the RTL-C level made possible by HDL-based tools and flows. These requirements included:

- Easy, rapid, low-risk architecture derivative design
- Risk reduction for complex product designs: system concepts must be rapidly implementable with little risk of architectural infeasibility
- Abstract models that are easy to generate
- An ability to make first-order trade-offs and architectural and IP evaluations above RTL level

P.J. Ashenden et al. (eds.), System-on-Chip Methodologies & Design Languages, 297–306.
© 2001 *Kluwer Academic Publishers. Printed in the Netherlands.*

- Real and effective HW-SW co-design and co-verification at a higher level: applications running on an architecture without the overhead of verifying complete details at the RTL level
- Easy to create, and modify on-chip communications schemes, and to change the interfaces of IP blocks
- Linkages to real HW and SW implementation flows

We now report two years of progress in developing design methodologies and technologies that have delivered function-architecture co-design methodologies and tools and seen them utilised in both research and industrial design projects. This methodology marries together the notions of platform-based SOC design [3] and an abstract design methodology into a complement of tools, which allow:

- semantic composition of application models drawn from various computational domains into a system behavioural model.
- capture of an abstract architecture, composed of custom design and reusable IP elements.
- explicit mapping of function to architecture,
- performance analysis,
- communications refinement
- generation of HW and SW implementation data.

The methodology has been delivered both in tool form, and as part of a more comprehensive System-On-Chip design methodology, to the Alba project in Scotland. The tools, which have been commercially released as Cadence's Virtual Component Co-design (VCC) [4], were developed with a number of systems, semiconductor, and IP partners, in the 'Felix' collaboration. We will discuss some of their practical applications of this technology for wireless, multimedia, automotive, and other products.

2.   FUNCTION-ARCHITECTURE CO-DESIGN

Function-architecture co-design [5] is a system level design methodology (Figure 1). In this modelling approach, designers describe a system in two orthogonal ways:

- The function that the system is intended to provide; i.e., a model of the application, and
- A candidate architecture which has the potential of realising the function, including choices of IP blocks or virtual components in both HW and SW domains.

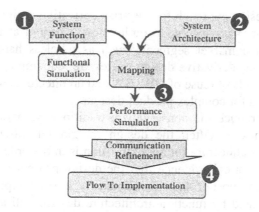

Figure 1. Function-Architecture Co-Design Methodology

The function is then explicitly mapped to the architecture, during which time all required hardware-software, hardware-hardware, and software-software partitioning is carried out. Both functional processing and communications interfaces are explicitly mapped to architectural resources. Performance analysis is used to study the suitability of the architecture and modifications are made until the design is brought into a feasible region. Detail is then added to the initial abstract models of inter-function communication, which refines it down to generic bus-transactions and protocol patterns, targeted to the chosen buses. Finally, export tools generate outputs that plug directly into HW-SW co-verification environments, and HW and SW implementation flows.

3. TECHNOLOGY AND REQUIREMENTS

Based on the requirements of 1998, we learned the following:
– Easy, rapid, low-risk architecture derivative design:
This was motivated by the realisation that many designs are based on previous generations of products, and in particular, the underlying product architecture is evolved with each product generation to incorporate new or revised functionality. As well as achieving key product goals for performance, power and cost, the flexibility of the architecture for new, 'derivative' products is vital. Complex, hierarchical, block-based SOC designs would be designed to fit a product niche in a cycle of 12-24 months. Derivative products might then take 5-9 months. In 1998, the concept of 'SOC integration platforms' was just beginning to be recognised as a growing trend. As of 2000, this trend has become ubiquitous [3].

A systems design approach for low-risk derivative design is necessary to avoid the low productivity and complexities of design at the RTL and C level [2]. In some market segments such as wireless handsets and other consumer products, derivative design cycles are reducing to periods of 8-12 weeks. High levels of reuse of HW, SW and architecture are thus essential.

– Risk reduction for complex product designs

We want to reduce the conservatism traditionally involved in systems design approaches. Often the design margin left after analysing the suitability of an architecture for an application is in the order of 50%, due to anecdotal and historical evidence that the design process at RTL level and below will 'eat up' most of this. Paradoxically, a better approach to systems design, as illustrated by function-architecture design, will allow the design margin to be reduced (e.g., to 30-40%) whilst at the same time reducing the risk that an abstract architecture will be impossible to implement.

– Abstract models that are easy to generate

New design approaches at the system level demand new modelling approaches. This is especially true of the complex set of models required for IP reuse - the number of models required at RTL level and below is very large. When the system level of design is considered the variety and complexity of models escalates. This will be further compounded with IP models of embedded software.

In order to ask IP providers and system modellers to create new abstract representations of their HW and SW blocks, it is essential to make those abstractions conceptually simple, and to support their creation with methodologies and detailed examples. The technologies utilised include software performance estimation based on a 'virtual instruction set' for micro-processors; 'kernel functions' for Digital Signal Processors (DSPs); a simple delay scripting language for dedicated HW and SW blocks; abstractions for real-time operating system latencies and services; and transaction delay models for buses and memories.

– An ability to make first-order trade-offs and architectural and IP
  evaluations above RTL level

This rests on the accuracy and field of use of the architectural components and their associated models. System level design models are approximate, with an error margin wider than at the RTL-level. However, characterisation of a platform and its IP blocks, together with prudence in drawing conclusions from performance analyses, can lead to satisfactory trade-off experiments and results.

Fundamental experience with architectural trade-offs using function-architecture co-design are based on dynamic performance analysis, and static power and cost analyses. Progress in dynamic power analysis as a support

for architectural decision-making will require additional development of analytical theory and associated models.
- Real and effective HW-SW co-design and co-verification at a higher
  level: applications running on an architecture without the overhead of
  verifying complete details at the RTL level

A discrete event simulator, integrating control-oriented and dataflow models, together with data-full, token and transactional-based models, can offer simulation speeds at least 100 times faster than the equivalent simulation at an RTL and C-level using cycle and pin-accurate models. Thus architectural trade-offs can be made with real application code.
- Easy to create, and modify on-chip communications schemes, and to
  change the interfaces of IP blocks

The notations used for communications modelling, at the abstract, token-based and generic bus-based levels must support an easy ability to change schemes and re-analyse. Interfaces should be abstractly described and thus simple to modify; communications refinements to particular patterns or templates must be easy to modify.
- Linkages to real HW and SW implementation flows

This requirement is fundamental. Too often in the past, system level tools and design has been the preserve of architectural 'gurus' who have not been able to share with implementers the underlying models and decision-making analyses which underpin their designs. As a result, implementers, when making micro-architectural decisions, have no common model base or medium with which to discuss their detailed designs with the system architects. The need to recapture the design during implementation has meant that system level models are often a dead-end. System-level design must be built into a comprehensive design flow from concept to implementation [6].

4.   SYSTEM MODELLING EXPERIENCES

4.1   Ericsson

Ericsson, an early partner, has been evaluating different architectural solutions for communications systems – primarily wireless. [7] Their experiences reinforce the need for a clear separation of functionality and architecture, to allow much easier modification of the design at the system level, and reuse of major design elements.   Explicit bindings between functions and components are important. The tools must make this both intuitive and easy to modify. The ability to make 'what-if' changes in architectures and mappings is important to build understanding of the system under design. HW-SW partitioning as an implicit part of the explicit mapping allows both pre-partitioning of parts of the design, and exploration of alternatives. Better estimation and analysis is desired, but Ericsson's experience is that "much can be achieved with very simple models and good 'guesses'. The important part is to capture the assumptions in a structured manner and make them explicit." [7, p. 210]

4.2   Philips

Philips used the co-design tools to import application models for video systems, built in a modelling tool called YAPI (Y-chart Application Programmers Interface) [8]. These used a lossless parallel process model of computation, Kahn Process Networks. A key requirement was to support several different computational models, or alternatively, allow model importation. The application modelled was an SOC decoder for digital video broadcast, involving about 60 IP blocks, including a MIPS processor and accelerating hardware. The tool was used to explore mappings and communications refinement. Communications latencies were modelled using simple delay equations for token-based transfers. At the aggregate system level over many transactions, good fidelity was expected for video applications. Benefits of using this approach included simulation speed – for 66 processes and 201 communication channels, one second of real-time video was simulated in less than 5 minutes. This permitted both effective IP selection and validation of algorithmic compliance with customer requirements. The clear separation of function and architecture, and function and communications, was an effective approach for defining an SOC integration platform, and allowed more system designers to work on derivative products.

4.3   Magneti-Marelli

Magneti-Marelli, a European automotive systems company, used the technology to model an Engine Control Unit (ECU) with about 100 functional blocks. [9] This system model included blocks for primary synchronisation, ignition and injection, the control chain and device drivers. In their experience, the software performance estimation technology provided good accuracy when compared to real values measured on an in-circuit emulator. This allowed them to compare different hardware platforms without relying on real development boards and complex code porting. They were able to define platform requirements more precisely, and optimise system implementation.

4.4   Thomson-CSF

Thomson-CSF communications modelled both a telecommunications SOC and a smart card SOC in the co-design technology [10]. The telecom SOC included 3 processors: a 5 GOPS SIMD processor, a 1 GFLOP DSP core and a 32-bit RISC core.   The model included the complex bus communications between the processors, both synchronous and asynchronous. The performance analysis simulations ran up to 100 times faster than RTL level. In the smart card SOC, which used legacy code, the designers did an architectural trade-off between an ARM core and reconfigurable hardware.   Design experience indicated effective use of design abstraction and the IP evaluation and trade-off capabilities. A key capability for future use is the links to HW and SW implementation.

4.5   Berkeley Wireless Research Centre: BWRC

BWRC is using this co-design technology as part of its Two-Chip Intercom (TCI) project [11], to support architectural exploration in the mapping of network protocols and baseband signal processing to software running on Tensilica configurable microprocessors, and dedicated HW blocks. A SONICS bus is used for communications. Exploration of different architectures and component choices has been done, including an analysis of ARM vs. the Tensilica processor, and a variety of mappings.   BWRC reported [12] that this approach for TCI allowed up to 4 different architectures and mappings to be explored per day, achieving convergence on a solution within 2-3 days.   This followed four person-months of development time, which included learning the technology, deciding on a modelling approach, and modelling the system.

5.   LESSONS LEARNED

5.1   Acceptance of function-architecture co-design

A very positive lesson, learned very early, was the natural acceptance by many partners and others of the essential and critical function-architecture orthogonal concept for systems design This is for most people a radical new design approach, avoiding the intertwining of system functional models with system implementation models.   However, to our surprise, this approach gained rapid acceptance by almost all designers to whom we talked.

5.2   SOC complexity

We originally anticipated relatively simple SOC complexity, composed of RISC micro-controllers, coupled with DSP processors for most designs. Very quickly, partners started to show us designs with 32-bit control processors, caches, multiple data buses, and extensive Direct Memory Access (DMA). They wanted to quickly explore the effects of radically changing the connection topology. We were challenged to enhance the fundamental modelling capability of the technology to handle the variety and complexity of the communication paths these architectures contained.

5.3   Communication design as a co-design problem

People commonly talk about the major system-design problem being HW/SW co-design. While this is important, one of greater import is to determine the communication strategy between behaviours mapped to the architecture, and the necessary timing and resource consumption. Designers want to make these decisions with confidence, before they incur the time, risk and expense of detailed design.   At the same time designers are determining their architecture, they have to consider the effects of choosing, for example, an interrupt-based method for communications versus polling. Evaluating the performance and resource implications of various communication strategies before the availability of an RTL co-verification simulation is very important. While supporting a fixed library of communication patterns is a good start, this is not sufficient. Specific behaviour and architecture lends itself to particular optimised communication patterns. Tools must allow the designer to easily customise and simulate patterns, and move them to implementation.

5.4   The importance of platform-based design

Originally we did not expect the SOC platform approach [3] to become as dominant as it has in recent years. This co-design approach is by no means limited to platform-based SOC. Nevertheless, the linkage between the two concepts [13] allows a profound leap in SOC design productivity. For non-platform based, hierarchical block-based SOC design based on reusable IP, the usefulness of the co-design is based on the availability of system models, and re-usability of system models over several generations of product design based on an initial architecture.

5.5   Role of software performance estimation

Originally we felt that designers would use software performance estimation to choose the microprocessors for their design.   However, IP providers are extremely reluctant to give appropriate abstract models, due to the restrictions on their appropriate 'field of use'.   Furthermore, choice of processors is often done on other criteria than raw performance: power consumption, software tool chain, business relationships, optimised legacy software, etc.   We now believe that the primary role of software estimation will be to measure the 'fit' of a set of application tasks on a processor, and to estimate the remaining design margin.   In addition, once the estimation technique has been characterised for a particular platform and application, it can continue to be used as a substitute for a cycle-accurate Instruction Set Simulator (ISS) in system trade-offs, at much greater speed.   Experience by some partners validates the accuracy of estimation for specific processors and applications.

5.6   New software modelling approaches and UML

The traditional worlds of software modelling and development, often using object-oriented approaches, and hardware design, have usually been separate.   The systems approach involved in function-architecture co-design of SOCs forces these worlds to collaborate.

One recent trend of considerable potential for the future is that of Object-Oriented Analysis, as represented by use of the Object Management Group's Unified Modelling Language (UML) [14], to be extended to embedded real-time software [15].   OMG has issued a Request for Proposals for extending UML via a profile covering aspects of scheduling, performance analysis and time, all vital for real-time software [16].   This RFP has attracted submissions (August 2000), and development of associated tools will occur over the next few years.   Further convergence between modelling of embedded real-time software, and co-design, is important.

5.7   Direct support for multiple models of computation

As seen in the Philips case, there are many models of computation of importance to system designers, and no industry/research consensus converging on just one or two.   Import strategies are useful to allow co-existence; in the long term it will be better to have native support for multiple models within one integrated technology.   This will clarify and make more rigorous the semantics of communication between these models.

6. CONCLUSIONS

We have reviewed two years of experience in developing and applying function-architecture co-design methodologies and technology. Partner experience indicates the merits of this approach in the design of IP-reuse dominated SOCs and systems. The next few years will see the long-awaited 'take-off' of system design as it is linked into the general design flows for complex systems. We believe that the function-architecture approach will be a fundamental component of this take-off. Further development of the technology, especially in modelling, estimation, and links to embedded software, will be important for this technology to become ubiquitous.

REFERENCES

1. G. Martin and B. Salefski, "Methodology and Technology for Design of Communications and Multimedia Products via System-Level IP Integration", Designer Track of DATE-98, Paris, France, February, 1998, pp. 11-18.
2. G. Martin, "Design Methodologies for System Level IP" ", DATE-98, Paris, France, February, 1998, pp. 286-289.
3. H. Chang, L. Cooke, M. Hunt, G. Martin, A. McNelly, L. Todd, Surviving the SOC Revolution: A guide to platform-based design, Kluwer, November, 1999.
4. M. Santarini, "Cadence rolls system-level design to fore", Electronic Engineering Times, January 10, 2000, pp. 1, 22, 24, 130.
5. S. Chakravarty and G. Martin, "A New Embedded System Design Flow Based on IP Integration", DATE-99 User Forum, pp. 99-103.
6. M. Baker, E. O'Brien-Strain, "Co-Design Made Real - Generating and Verifying Complete System Hardware and Software Implementations", Embedded Systems Conference, San Jose, September 1999.
7. J. Plantin, E. Stoy, "Aspects on System Level Design", CODES-99, pp. 209-210.
8. J-Y. Brunel, E.A. de Kock, W.M. Kruijtzer, H.J.H.N. Kenter, W.J.M. Smits, "Communication Refinement in Video Systems on Chip", CODES-99, pp. 142-146.
9. "Magneti-Marelli: Cierto Virtual Component Codesign (VCC) environment for engine control unit design", available at www.cadence.com.
10. "Thomson-CSF: Cierto VCC for system-level design of telecommunications and smart card SOCs", available at www.cadence.com.
11. J. Ammer and J. da Silva, Jr., et. al. "Design Flow for Wireless Protocols", BWRC presentation, January 2000, available at bwrc.eecs.berkeley.edu/Research.
12. J. Rabaey, A. Sangiovanni-Vincentelli, and R. Brodersen, "Communication/Component-Based Design and the PicoRadio Design Driver", GSRC review, December 9, 1999.
13. G. Martin, "Productivity in VC Reuse: Linking SOC platforms to abstract systems design methodology", FDL 99, pp. 313-322.
14. G. Booch, J. Rumbaugh, I. Jacobson, The Unified Modeling Language Reference Manual, Addison-Wesley, 1998.
15. B. Selic, "Turning clockwise: using UML in the real-time domain," Communications of the ACM, vol. 42, number 10, pp. 46-54, October 1999.
16. OMG UML Profile for Scheduling, Performance and Time RFP, OMG: ad/99-03-13.

Virtual Component Reuse and Qualification for Digital and Analogue Design

Natividad Martínez Madrid and Ralf Seepold

FZI Karlsruhe, Haid-und-Neu-Str. 10-14, 76131 Karlsruhe, Germany
martinez@fzi.de, seepold@fzi.de

Key words: Intellectual Property (IP), virtual component reuse, platform-based design, analogue and mixed-signal design, IP qualification

Abstract: This chapter gives an overview on past and future directions of applied research and application-oriented development in the domain of Intellectual Property (IP) reuse. Since a gap has to be bridged between available technology and methodologies and tools that can manage complex designs within a reasonable time frame, a new quality within IP has to be defined, thus enabling platform-oriented reuse for IP users and IP providers. Furthermore, the application of design reuse methodologies to analogue and mixed-signal components for System-on-Chip (SoC) is an emerging and revolutionary field. This article presents a summary of the current status in research and industrial application, together with information about leading standardization initiatives and future development lines in the area. Finally, the basis for an IP qualification flow is derived from a European initiative that is focused on System-on-Chip design. Taking into account the viewpoints of an IP user and an IP provider, a methodological approach is presented.

1.   INTRODUCTION

System design, and especially System-on-Chip (SoC) development, is one of the most important application areas in future-oriented designs, and design reuse methodologies are one key to introduce a new "quality" in design and design process [1]. To accomplish this, Intellectual Property (IP) management requires new concepts and an innovative break-through to

P.J. Ashenden et al. (eds.), System-on-Chip Methodologies & Design Languages, 307–316.

introduce a new quality in Electronic Design Automation (EDA). These concepts will enable enhanced, platform-oriented design flows from system level to layout level.

The huge increase of available integration capability in microelectronic chips during the recent years has lead to a paradigmatic change in electronic design. The challenge is no longer how to improve board-based design, but how to develop SoC for specific platforms [2]. Today's mixed-signal circuits constitute an important part of European and also worldwide integrated circuits. Due to the fact that many applications in industry have analogue front-ends but often need digital processing backends, analogue and mixed-signal devices are required to be more and more flexible and programmable.

Furthermore, complex environments require a drastic change of design methodology. In digital design, a major movement has been undertaken in the last years by inventing design reuse of Intellectual Property (IP) modules.

Analogue and mixed-signal design has turned out to be a hot topic in the industry and academia, mainly due to the fact that these analogue and mixed-signal IPs need to become part of the SoC, since Time-To-Market (TTM) remains the most critical factor for success. Analogue and mixed-signal virtual components (VCs, i.e. IP blocks) are needed not only in traditional analogue applications but also in "digital" ones. They will be required in high-performance, high frequency custom mixed signal integrated circuits such as partial-response, maximum-likelihood disk-drive ICs or on single-chip radios.

Several research approaches show first results, treating aspects like specification, validation, parameterisation, synthesis, circuit sizing and integration in the design. The next section sketches some of approaches from the analogue domain. It is followed in Section 3 by a description of a large digital reuse project (EURIPIDES) developed in the frame of the MEDEA (Micro-Electronics Development for European Applications) Programme. More information on the MEDEA EDA Roadmap is given in Section 4. Section 5 presents new research lines in the area of IP platform-based design and qualification.

2.   ANALOGUE DOMAIN

Analogue and mixed signal components present a huge amount of added problems when facing reuse of designs compared to digital circuits, or generally speaking, when comparing the automation and efficiency of the design process. Experts refer to an "analogue/digital design effort gap" in design [3]. This expression reflects the fact that, whilst analogue will take less than 20% of the circuit area of a new IC or system, the design effort required can easily consume 80% of the allocated time. The main reason for such a gap is that most parts of the analogue design are developed manually.

The answers that research groups in the academia, and also in the industry, are providing to bridge this gap are based on the following concepts.

New top-down design methodologies are needed to raise the level of abstraction of analogue and mixed-signal design. The designs should begin with system-level specifications based on behavioural descriptions of the components. A bottom-up verification strategy is applied. The models get adapted to the different design points and they are the basis for a process of cell characterization in which the design parameters are extracted. The information obtained is back annotated in the behavioural model, so that they can provide more realistic results in the system simulation.

One of the approaches is being developed in the FhG Dresden IIS/EAS with the collaboration of Infineon [4]. They have developed a modelling environment in the frame of a library of parameterised analogue and mixed-signal basic blocks. The library is organized in classes (for instance operational amplifiers). For each class there are different structures available (for instance, low-noise amplifiers, or high-bandwidth amplifiers). Together with the structures, the library must provide an infrastructure to facilitate an efficient redimensioning of the selected structure. To support design reuse, the library components should be parameterised at all levels.

The models for the top-down design are classical behavioural models, and for the bottom-up verification, a combination of numerical and behavioural models has been used. In this way, modelling can be treated as a systematic and almost automatic task. Besides the numeric model generation, the environment generates the corresponding documentation.

Other research activities in the area of analogue design automation and reuse are undergoing at the University of Kaiserslautern, together with Robert Bosch GmbH [5]. The approach followed in this case is a top-down, specification-driven design based on the use of macromodels and refinement of these through predefined strategies. Macromodels using ideal network elements are a well-known technique that allows a better insight in the behaviour of the circuit. A library of such macromodels, each of them associated with different circuit structures, facilitates also the exploration of design alternatives. In this case, analogue design reuse is supported by the creation and application of a reuse database containing this library, together with other design knowledge.

A top-down design guideline tool supports the design process. The designer starts with filling up the set-up requirements catalogue. In the next step, the system-level model is designed in a block design. Finally, the

310

macro-level model is extracted. In the next step of the design process, a mixed-model level description is obtained, in which the controlled sources have already been substituted by one of the associated structures equivalent. Through a tool and library supported architecture exploration process, all the macros are substituted by pin-compatible structures until reaching the device model. The design can be simulated at every stage of the mixed hierarchy and allows focusing on critical circuit parts. In this approach, the technology switch is not difficult because the mathematical formulas are generic.

The area of analogue IP cataloguing and documentation generation is also drawing a lot of attention. This is due to the fact that there is a huge amount of information associated to each analogue and mixed-signal IP. Not only the models are necessary, but also valid values for the parameters, working conditions, test and verification information, etc. One of the problems in automating and reusing analogue designs is that it is a very creative process and a lot of the information necessary to understand and apply/adapt the design resides in the designer's head. Therefore, the designer should try to produce an exhaustive documentation. The tools should support this task. An example of this approach is the DREAM system [6], developed at TEMIC Semiconductor GmbH. It is an HTML-based system where the designers can access the documentation of the components with links to tools for representation and re-entry.

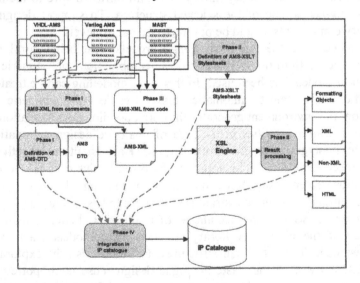

Figure 1. Characterisation and documentation for AMS reuse

A further step in this direction is being done in a common project among FZI, the University of Tübingen and Robert Bosch GmbH. The objective of the project is to create a common XML-based documentation system for analogue and mixed-signal virtual components expressed in an analogue description language (see *Figure 1*). In the current state, MAST and VHDL-AMS are being considered, but a front-end for Verilog-A/MS is also planned. The system should parse the designs to extract valuable reuse information about interface, parameters, internal structure and behaviour. The designer's comments are incorporated in the system, together with additional information of testbenches, tool configurations, etc. An extension to XML called AMS-XML has been defined to represent this information.

3. DIGITAL DOMAIN

MEDEA is part of the pan-European EUREKA network for industrial R&D [EUREKA Project S 1535]. Since its start in January 1997, MEDEA has organised and promoted the pre-competitive R&D co-operation of companies and research institutes among several European countries. As an example of an IP-related project that contributed to the MEDEA objectives, the project EURIPIDES[1] has provided specialized designs and design tools, resulting in design methodologies, to efficiently support the construction of complex systems in key application domains [7]. EURIPIDES conforms to one of the paramount aims of MEDEA [8], which is to maintain the leading edge that Europe possesses in the fields of telecommunications, automotive and multimedia technologies. The first two years (1997-98) of the EURIPIDES project have shown that the protection, creation and reuse of the European intellectual property continue to be a matter of primary importance.

From the partners point of view there are two pivotal solutions to combat the design gap. Firstly, design reuse in conjunction with a migration to higher levels of abstraction will augment the productivity by a factor of 5 to 10, and secondly, prototyping will decrement the time to market.

The main thrust of EURIPIDES is to facilitate the reuse of digital IP in a real industrial environment, as this offers the greatest potential for cutting design costs and cycle times. With the number of transistors on a typical chip doubling each year, complete designs from one generation can now be used as building blocks in even more complex integrated circuits (ICs) and systems. From recent results, it is estimated that reusing standard blocks can slash development costs by 50 to 75% and halve the time to market. With a more flexible approach whereby blocks can be customized for specific applications through reuse of the underlying design knowledge, the savings may reach 85% – with development cycles becoming three to eight times

---

[1] EURopean Intellectual Property In Designing Electronic Systems; funded as SSE-Project by the German Ministry BMBF and labelled from MEDEA as A-407.

shorter. The urgent tasks for future development in this area will remain in the scope of new activities that are compliant to concurrent development shown in *Figure 2* [9].

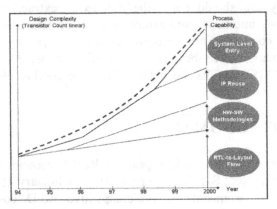

Figure 2. Process design gap

The basis for a European strategy on automation and IP reuse can be adopted by organizations irrespective of their own particular tool preferences and design flows. This will be promoted in discussions with the Virtual Socket Interface Alliance (VSIA [10]), the body responsible for setting world standards for mixing and matching 'virtual components' from multiple sources, to which Europe has hitherto made only a modest contribution.

Participation in MEDEA has been particularly valuable in allowing a fruitful exchange of information and facilities with other project teams working in related areas. A longer-term aim will be to unify the approach to design automation in both digital and analogue circuitry. Currently, a follow-up project has been defined and submitted to MEDEA+ that will attack open issues and new methodologies urgently required to push reuse application in digital and analogue areas.

4.   THE MEDEA EDA ROADMAP

Driven by the specific needs of European industry and derived from several experiences that had been taken into account during the recent four years of MEDEA, an initiative was started to establish a view for a MEDEA EDA roadmap that will also have impact on several areas relevant for MEDEA+[2] [11]. The main challenge is facing two domains: development of

---

[2] The follow-up of this initiative has been already decided and labelled as the MEDEA+ programme (EUREKA project (E!2365)).

design automation technology to cover design feasibility up to specification and reuse of design, and optimisation of automated design solutions to reach break-through in design cycle time by providing application platforms [12].

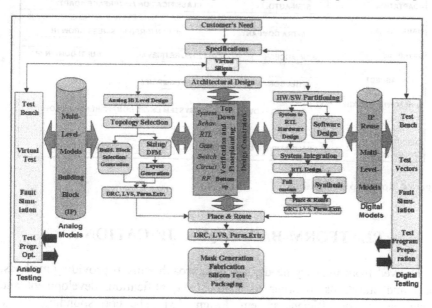

Figure 3. Design Flow

The general architecture of the mixed analogue/digital design and test flow is shown in *Figure 3*. This shows the backbone of the digital flow from specifications validation with the customer until final packaging.

With respect to the challenge to invent IP reuse with multi-level models, current approaches are mostly restricted to certain levels of abstraction, technologies, design flows, tools or even to a different business model that the one applied. Efficient reuse needs to follow a four dimensional approach this is facing hardware, software, digital and analogue (mixed-signal) design. None of the available tools can manage these dimensions, and furthermore, none of the existing tools can handle the complexity of the next or even the current chip generation.

Several key topics are selected for short and long-term evolution forecast: exchange, adaption, management, cataloguing, legal and business aspects, integration in the design flow and specific applications (see *Figure 4*). However, design quality seems to be of overall relevance. This covers the correctness of a product as well as estimation for support and costs during the development and before the integration.

	00	01	02	03	04	05
EXCHANGE	IP VIRTUAL LIB.		PLUG & PLAY LIB.		IP CREATION	
ADAPTATION	SEMI-AUTO.		CLASSIFICATION / INTERFACE ADAPT.			
MANAGEMENT	INTRA COMPANY		IP CENTERED BUSINESS GROWTH			
DATABASE	MAN. ACCESS		HIGH LEVEL RETRIEVAL		MULTI-DOMAIN IP	
LEGAL ASPECT	INTRA COMPANY		INTER COMPANY			
LINK TO THE DESIGN FLOW	EXISTING FLOW		SYSTEM LEVEL COMPREHENSIVE FLOW			
SPECIFIC APPLICATIONS	VALIDATION IN MEDEA PROJECTS			EUROPEAN NETWORK		

Figure 4. Roadmap for IP Reuse

5.   PLATFORM-BASED QUALIFICATION

Derived from industry needs, a new approach aims at providing methods, tools and standards in order to enable IP qualification, development and reengineering for efficient system design [13]. The corresponding design flow will contain methods and tools for IP entrance checks customisable to the IP users needs and relevant to IP vendors. Moreover, this design flow will also consider methods and tools to support distributed design teams that facilitate inter and intra-company co-operation. Flexible parameterisation of qualified IP will provide application-specific customisation by proposing and supporting standardized formats.

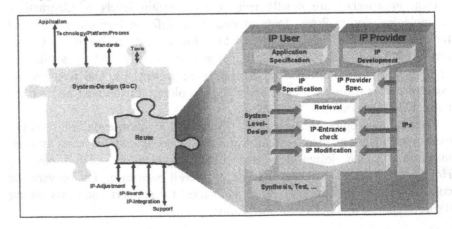

Figure 5. IP qualification design flow

Design Flow Starting at the problem definition and ending up at the IP application, the design flow proposed will offer several interrelationships (see *Figure 5*). Both the point of view of the IP providers and that of the IP user/integrator will be incorporated, driven by concrete industrial applications.

In the core of the IP qualification design flow, IP user and IP provider are interrelated via specification, retrieval, IP entrance-check and modification. In this area, our main research effort will contribute to the following aspects:

1. *IP Specification*

   The initial step is the elaboration of a requirements document to be used as a reference. Specification languages and guidelines both for digital and analogue components need to be taken into account. From these specification concepts, the necessary attributes of virtual components will be derived.

2. *Design Guidelines and Standards*

   Design guidelines for IP authoring have already been successfully applied and have become a de-facto standard. The work of FZI will be to study and define IP quality features to be contributed to standardization initiatives.

3. *Retrieval*

   An existing Reuse Management System [14], which allows intelligent retrieval of IP, will be refined to improve metric calculation, security aspects and extended to support a common XML-based documentation system for analogue and mixed-signal virtual components (see *Figure 1*).

4. *IP Entrance Check*

   The main research activity will be the definition and extraction of quality measurements to evaluate and compare IPs.

5. *IP Modification*

   The goal to be fulfilled in this part is the definition of a method for modification mechanisms to be able to adapt a (parameterisable) component to particular working environments.

6.   CONCLUSION

Besides the definition of efficient and innovative methods and tools, the European microelectronic industry needs to contribute to international standardization efforts. The extension of design reuse methodologies to integrate the creation, exchange and application of analogue and mixed-signal virtual components (IP) is a new and promising field that has attracted the interest of industry. IP qualification is a prerequisite for the realization of

IP reuse for multi-level models. Those models are inherently belonging to the MEDEA design flow and thus they remain part of the core for an efficient solution for IP reuse. The developed design flow and the accompanying proposed actions will form the platform to establish several qualification attributes and measurement facilities able to support comprehensive reuse and on-time product design for emerging markets like telecommunication or automotive.

REFERENCES

[1]   R. Seepold, "Standardization of System-Level IP", in Proc. of the Workshop "Methoden und Beschreibungssprachen zur Modellierung und Verifikation von Schaltungen und Systemen", Meißen , Germany, 19-21. Feb. 2001.

[2]   H. Chang, L. Cooke, M. Hunt, G. Martin, A. McNelly and L. Todd, "Surviving the SOC Revolution. A Guide to Platform-Based Design", Kluwer Academic Publishers, 1999.

[3]   N. Martínez Madrid, E. Peralías, A. Acosta and A. Rueda, "Analog/Mixed-Signal IP Modeling for Design Reuse", in Proc. of DATE Conference, Munich, Germany, 13-16 March 2001.

[4]   J. Eckmüller, R. Jancke, A. Schwaferts, P. Schwarz, P. Trappe, "Verhaltensmodellierung und Bibliothekskonzept für analoge Grundschaltungen der Telekommunikation", in Proc. of the Analog'99 Conference, Munich, Germany, 18-19 Feb. 1999, pp. 40-45.

[5]   P. Jores, R. Sommer, G. Gielen, "New Methods in Analog System Design", in Proc. of the MEDEA/ESPRIT Conference on System-level design, Antwerp, Belgium, 22-24 Sept. 1999.

[6]   V. Meyer zu Bexten, A. Stürmer, "Analog Design Reuse Information and Application System", in Proc. of the MEDEA/ESPRIT Conference on System-level design, Antwerp, Belgium, 22-24 Sept. 1999.

[7]   EURIPIDES, EURopean Intellectual Property In Designing Electronic Systems, http://euripides.fzi.de.

[8]   MEDEA, Micro-Electronics Development for European Applications, http://www.medea.org.

[9]   EURIPIDES (A-407) Project Profile, http://www.medea.org.

[10]   VSI, Virtual Socket Interface Alliance, http://www.vsi.org.

[11]   MEDEA, The MEDEA DESIGN AUTOMATION ROADMAP, 2nd release, http://www.medea.org/doc/page_doc.htm, 2000.

[12]   R. Seepold and N. Martínez Madrid (eds.), "Virtual Components Design and Reuse", Kluwer Academic Publishers, ISBN 0-7923-7261-1, 2001.

[13]   A. Vörg, N. Martínez Madrid, R. Seepold and W. Rosenstiel, "IP-Qualifizierung wiederverwendbarer Schaltungsbeschreibungen", in Proc. of 10. E.I.S.-Workshop, Dresden, Germany, 3-5 Apr. 2001.

[14]   R. Seepold and N. Faulhaber. "A Flexible classification model for reuse of virtual components", in Reuse Techniques for VLSI Designs, Ed: R. Seepold and A. Kunzmann., Kluwer Academic Publishers, ISBN 0-7923-8476-8, 1999.

INTERFACE BASED DESIGN

Using the VSI System-Level Interface Behavioral Documentation Standard.

Brian Bailey,   Gjalt de Jong,   Patrick Schaumont,   Chris Lennard
Mentor Graphics, Telelogic, *   *IMEC,*   *Cadence Design*

* Was employed by Alcatel when this was written

1. INTRODUCTION

In April 2000, the first version of the Virtual Socket Interface alliance (VSIA) specification entitled "System-Level Interface Behavioral Documentation Standard"[1][2] was published and received wide acclaim by academic, industrial and press people [3]. The specification, which outlines a methodology for the separation of behavior from interface, provides a mechanism for documenting the intent and implementation of interfaces for Virtual Components (VCs) as well as a mechanism for showing how the interfaces at various levels of abstraction relate to each other. This specification results in a number of advantages to all participants in the design process.

Firstly, it improves rapid comprehension of the VC itself by allowing system integrators to see the 'functional' interface into a component without needing to look at the details. This functional interface

317

P.J. Ashenden et al. (eds.), System-on-Chip Methodologies & Design Languages, 317–331.
© 2001 *Kluwer Academic Publishers. Printed in the Netherlands.*

provides an abstraction that allows Hardware and Software engineers to effectively communicate with each avoiding many of the troublesome problems that are only found during system integration and test.

Secondly it assists in the creation of the VC itself by allowing comprehension of the intent of integration independent from its final utilization.

Thirdly, the mapping between the levels of abstraction helps alleviate integration and verification problems since a path is provided to show why interface details are necessary at all levels of abstraction. This has important implications for the verification flow since it becomes possible to write abstract test benches that are capable of driving detailed design descriptions or for encapsulating a detailed part of the design within a high level model of its surrounding environment.

Finally, by describing the interface separate from the behavior of the VC, it provides protection for the component provider by allowing a high level behavioral model to be used without exposing any of it's internal function.

This chapter will provide an in depth look at the standard, which has been demonstrated by two pilots that have been conducted, one by Alcatel[14] and the other by IMEC[4].

2. HIGH LEVEL CONCEPTS

Commonly, system architects begin considering their problem from the perspective of functional or behavioral task definition [5]. These tasks are linked by "ideal" channels, through which information is sent and received as needed, without concern for any conflicting resource requests. There is no need for such concern at this stage, as the architecture is dealing with functionality only and not communication protocols. As this design is refined, common communication resources are specified, control protocols administered, and sharing of functional units identified. The common issues associated with system design become visible and the design moves from that of the ideal Figure 1, to that of the realisable, Figure 2. Although at the finest level of detail, the design may appear to have changed relative to high-level conceptual models, the fundamental operational principles must be inherited. The implemented, or "realized", tasks of Figure 2 must perform the conceptual tasks of Figure 1. In fact, the model should remain completely valid throughout the entire design refinement process.

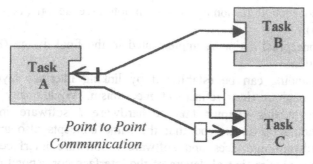

Figure 1. Conceptual View

This refinement can be viewed as one of tightening the design envelope until the design space converges to an implementation. The implementation involves greater complexity and dependence upon the actual channels of communication used to transfer information within the design.

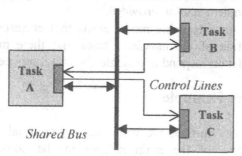

Figure 2 Refined Design

3. INTERFACE LAYERING PRINCIPLE

The SLIF standard introduces the concept of Interface Layering, or Abstraction Hierarchy. An interface layer is a translation wrapper that is capable of taking one level of interface abstraction to the next more detailed level. For example, if a Virtual Component (VC) behavior uses the conceptual communication channel of a "blocking read" at the top-most abstraction layer, one layer of the refinement below this can refine the "blocking read" as a read action dependent upon an asynchronous "Request / Acknowledge" action. A further layer below this, the "Request / Acknowledge" actions could be tied to synchronous clocking actions in a final HW implementation. In each case, it is critical to understand:

- What the conceptual reasoning is for an interface action (Top Down Justification).
- How the conceptual action is implemented in the final design (Bottom Up Verification).

This understanding can be established by linking interface layers, or abstractions, through a clear refinement mechanism. While many of the examples given in this chapter take a hardware / software interface viewpoint, it should be understood that the same concepts also apply to hardware / hardware interfaces and software / software interfaces. The standard does not require that all layers of the interface correspond directly to the implemented interfaces and allows for some of the layers to exist only in a documentation form. Three types of layers are recognized, namely:

- Documentation – the interface is completely described as well as its relationship to the layers above and below it.
- Executable – a model of the interface at this level is provided although there is no direct implementation provided.
- Implementation – An executable model exists that corresponds directly to an implementation of the interface. Note that there may be many implementations that correspond to a single abstract interface.

3.1 Conceptual Example

Before going any further, we will give a conceptual example that helps to illustrate some of the main notions in the decomposition of communications. In Figure 3 we show a high level definition of the communication. While the exact diagrammatic terminology will be defined later in the section on *interface behavior*, it should still be possible to understand the intended concepts.

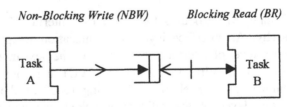

Figure 3. Interface Behavior

What we see in Figure 3 are two functional tasks, represented as Task A and B, which are connected by a FIFO of infinite depth. Task A writes data into the FIFO and task B reads from it. Task B can attempt to do a read at any time and after it makes a read request will be blocked until the data is returned. At this point no assumptions are made about how this is to be implemented and clearly the notion of an infinite depth FIFO will need to be resolved. At the next level of refinement, shown in Figure 4, it has been resolved that communications will be done using bus based principles.

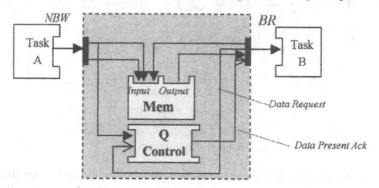

Figure 4. Layer 0.C Channels and Interfaces

In Figure 4, and the following figures, the black boxes represent transformations that are made to resolve the higher level communications down to a lower level. These clearly need to be fully described in order to understand how this refined system operates. Note however, that it still does not imply that the communications channels are shared via a common bus. The abstract notion of the FIFO has been broken down into two tasks, one memory block and a separate control unit. The diagram does not show the communications between the control unit and the memory to avoid too much complexity. If task A is now mapped to software we could assume a further stage of refinement as shown in Figure 5. Here we see that a driver is now controlling the FIFO operations, which is part of the Operating System. Finally we might decide that both sides of the communications would share a common bus, which is implemented using the VSIA On Chip Bus transaction layer [6].

This level has not been shown in its entirety, instead, we show in Figure 6, the details of Task A from its abstract, Layer 1.0 interface to its Layer 0.0 protocol specific, or implementation specific, interface. This shows the virtual components with which it shares an immediate connection – namely the Operating System and the Driver.

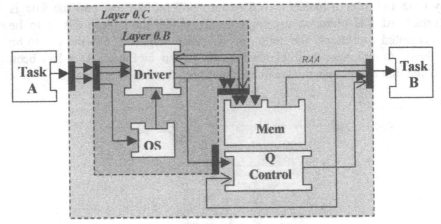

Figure 5. Layer 0.B Software Layer

This exercise clearly demonstrates how much simpler and more intuitive it is to start describing the interfaces at a high level. It is also very difficult once the interfaces have become fully expanded to know what the original intent was and thus makes it very dangerous to make changes.

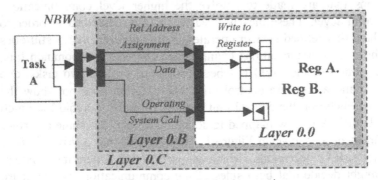

Figure 6. Layer 0.0 Implementation

4. INTERFACES IN THE CONTEXT OF SYSTEM LEVEL LANGUAGE DEVELOPMENT

There has been a lot of effort recently to define new languages for system level design. Some of these efforts have come out of the software world, such as SDL and UML, while the existing language committees such as OVI and VI (now merged into a new group called Accellera) have worked on the update or definition of new languages for hardware design. In addition, new work is going on in a number of academic institutions such as UC Berkeley, and a number of new language organisations have been formed such as SpecC and SystemC.

In all of these other efforts, the focus is on the definition of functionality, such as would be found in Tasks A, B and C in Figures 1 and 2 above. The expectation is that functional tasks communicate with each other through the primitive mechanisms provided within the applicable model of computation or through conversion facilities available in the backplane. In other words they do not provide a methodology that enables the separation of the definition of the interfaces from the functional tasks. This lack of functional separation makes it difficult to do project or geographic partitioning since there is no way to formalize the protocol of the interface and this makes the incorporation of IP almost impossible. This says that the work of the VSIA system level interfaces (SLIF) group has two very important roles to play.

- The first is that it should provide a set of guiding principles to these other system level language efforts. This will ensure that the semantics they define are sufficiently rich to be able to support all, or a useable subset of, the communications defined as part of the SLIF methodology. Thus while SLIF does not define a languages it can be viewed as providing a dictionary that needs to be supported by the system level languages.
- Secondly, SLIF provides a methodology for using such languages as will be described in the later sections of this chapter. It is the combination of language and methodology that will allow the effective use of system level design methods.

5. STRUCTURE OF THE SLIF DOCUMENT

The document requires that at each level of interface abstraction, that all aspects of the interface be described along with how these interfaces relate to the level of abstraction above them. At the highest level, called Layer 1.0, the overall communications intent of the component should be made clear. It should be described in terms that do not consider mapping into either hardware or software. At the other end of the scale, called Layer 0.0, the interface is fully mapped. It gives interface properties at the RTL-equivalent level for both hardware and software components. At each level, the interface is described in a number of sections, namely the data types used, the ports and their attributes, the interface behavior and the transactions used. In addition the association between layers includes the data type mapping or coercion, the port mapping and the behavioral mapping. In this chapter we concentrate more on the definition of the layers rather than the inter-layer mapping. Each of the attributes of the layer will be described briefly.

5.1　Data Types

The VSIA data-type standard addresses the issue of highly variable system-level data-types and operations. This variability in the industry has made the portability of software VCs and simulation models difficult. The solution offered by the VSIA Data-Types standard is one of standard header-file syntax and data-type semantics for C/C++-based data-types. The goal is to achieve a result similar to what the IEEE 1164 [7] package has achieved for VHDL. Standardisation of data-types is recognised as critical in any system-level C/C++ language. Example language developments that have identified this need are SystemC [8], Cynapps [9], Superlog [10], C-level Design [11], and Frontier Design [12]. The VSIA is seeking data-type alignment will all these efforts

The current VSIA data types specification specifies C/C++ header files. The data types and their semantics can be carried over to any (underlying) language for system level specifications. Accellera will also adopt the VSIA datatypes in the architectural language committee and in their 'C' working group [13].

5.1.1　Basic Concepts

The data types and their semantics are based upon the following concepts:

- no assumptions on compilers and run-time environments.
- explicit conversion functions to avoid implicit conversions.
- all types have a string representation to serve as foreign exchange mechanism thereby eliminating reliance upon stdio or IO stream class libraries.

- defined initial values for all types and implementations to have freedom with respect to data representation, additional C++ keywords, type inheritance, analysis capabilities, etc.

The group has identified the need for the development of a test-bench set so that vendor compliance can be measured.

For consistency between all related types and their semantics, a user centric use model is applied. This provides a rich set of functions and operators. For performance, good coding style of the defined data types is demanded. We classify the set of types into four classes: base-types, bit-vectors, signed/unsigned numeric types, and fixed-point types.

5.2 Port Definitions

A port is the connection point into a VC. It is the only way that information can flow into or out of a component. Communications between two functional tasks is accomplished through channels that connect ports together. There are a number of static attributes that ports have. At the higher levels of abstraction it is not sufficient to talk about ports as inputs or outputs since a port is capable of handling complex protocols that may have many components. Instead they are more generally described as a *producer* or *consumer* of data, or an *initiator* or *responder* for control information. It should be noted that *initiator* and *responder* do not directly map into a master/slave relationship, as only the commencement of action is identified in the later terms.

When an interface is refined, the SLIF document requires that the relationships between the ports at the higher level and those that exist at the lower level be described. This shows the structural decomposition that is going on in the design process. In addition, it also requires that the method in which the refined ports implement the behavior associated with the more abstract port be described. This shows the behavioral decomposition.

5.3 Interface Behavior

All activity through a port of a top-level description is referred to as a transaction. As the interface is refined, these transactions are refined into messages such that at the lowest level only messages remain. The difference between transactions and messages is their atomicity. This means that a message, which is atomic, either completes fully or not at all. No interference or interruption is allowed. Examples of messages may be the setting of a signal to a particular value. Transactions and messages have associated with them a number of attributes that help define its basic behavior, such as if values are persistent or only available at that instant of time.

5.4   Attributes

Attributes can be divided into three primary types. These are data attributes, control attributes and sequence or timing attributes. However, not all attributes are available at all levels of abstraction. For example, at the highest level, timing attributes cannot be assigned at all, whereas at the lowest level this may be essential. In between, there may be a need to add approximate or estimated timing.

Data attributes fall into two main areas. The first is direction, which defines the direction that data flows in. There is no typing of data on the interface itself, since this is derived from the port or protocol information. Neither does it imply the flow of a single type of data, or in the extreme that the data itself actually exist. For a performance model it may be sufficient to define that so much data of a particular type will flow, but the actual data not be defined. By itself, it is not possible to know when the data will flow, but must be combined with control attributes or a protocol for full definition. The second type of attribute is persistence. Without persistence the data is available only at the exact instance of time when the data is produced. Persistence can be provided by the addition of buffering, by a FIFO, or at lower levels of abstraction by a pipeline. FIFO's can be considered to be of infinite size at the higher levels of abstraction and of specified size as the design is refined.

A flow of control causes actions to happen in the behavioral tasks to which they are connected. Unlike the dataflow described above, there is no data contained in a control flow, but can be tightly associated with the flow of data. The basic form of control is non-blocking, which means that as soon as the message has been passed, the initiator will not be blocked by the execution of the receiver. This maximizes the parallelism in the system, which is essential in the high level design of a system. Combining control and data flows enables fully synchronous flow of communications. However, in some cases, it is necessary to modify a control attribute so that the initiator does block until released by the responder. This is accomplished by adding a blocking attribute. Other attributes necessary in some systems are notions of priority and timing or protocols. Protocols describe the exact sequence of operations that happen on a collection of flows. For example at the highest level we may have defined a blocking flow with a data transfer. When refined, this blocking flow may be implemented with a fully synchronous handshake using dedicated control signals and the data may be transfer over a bus. This was the case with our conceptual example shown in figure 4 for the flow into Task B.

5.5   Transactions / Messages

Attributes define the general characteristics of the channel and how the functional tasks respond to events that occur on the channel. Those events are defined by a set of transactions and messages. It should be realized that these are not fully independent and SLIF defines the appropriate transactions and messages that may correspond to each flow type as defined by the attributes. At the present time, SLIF defines the general form of transactions but does not define an actual API that corresponds to them. This is one direction that the work will take in the future.

At the highest level of abstraction, only a few of the possible transactions and messages are possible. This set includes *transRead* and *transWrite* for dataflows and *messSense* and *messEmit* for control flows. Even these have a number of restrictions on them. As the design is refined a much richer set of transactions and messages becomes available for use. In addition to the ones mentioned above, the interface may also use *transOpenChannel, transCloseChannel, transSynchronize, transReset, transControl, messRead* and *messWrite*. What may be counter-intuitive is that by the time the lowest level of abstraction has been reached, the available message set again becomes very restricted. No transactions of any type are permitted. The reason for this is that once an interface has been fully defined, all protocols should have been broken down into discrete signals. Thus all messages become either the transmitting of a piece of data, or the detection of any edge of data for control. This is what designers are actually used to seeing in the form of timing diagrams, or waveforms. It should thus be clear that without the interlayer mapping across all parts of the interface including the transactions that it becomes possible to discern the true meaning of the communications between two tasks.

6. EXAMPLES

The usage and benefits of SLIF will be demonstrated by two very different examples

The first example [14] looks at a large system level design problem conducted by Alcatel. As a systems company, Alcatel needs interoperable environments in which the various VC models can be linked at different abstraction levels. Existing VC encapsulation is insufficient to permit proper architectural exploration. Legacy (HW and SW) VCs are already implementation specific and, due to lack of proper system-level representations, it is difficult to integrate these in the top-down design flow.

In SoC design development, current best practice shows the reuse of interfaces in new designs. To move to common interface specification and re-use, Alcatel assessed the On-Chip Bus VCI and SLIF documentation specifications. The completeness and soundness of these specifications have

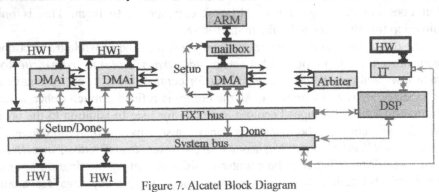

Figure 7. Alcatel Block Diagram

been checked with respect to the various interfaces that are in use and (re-) use within Alcatel designs, such as on-chip busses, and external interfaces such as Utopia for ATM; see Figure 7. In addition, the suitability of the OCB VCI is checked for use in multi-processor platforms.

Alcatel has defined an interface based design and refinement methodology shown in Figure 8. Functional and architectural elements are captured at different (interface and functional) abstraction levels. All component models and interfaces have various views, such as data, configuration, status and test. Transactions are defined at the highest level. These are refined to individual ports and signals and protocols at lower levels. The transaction interfaces to SW are used in top-level and system integration verification and testing. The lower level interfaces and protocols are verified once and used during module verification and test. This shows that interface based design complies very nicely with our test and verification strategies and methodology.

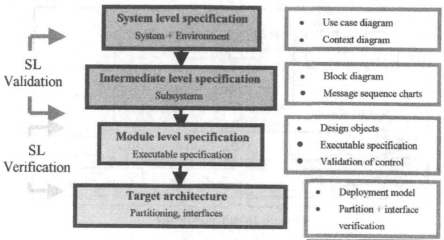

Figure 8. Design Methodology

The second example[15] is a much smaller but more detailed example that was actually used to validate some of the concepts contained within SLIF.

At IMEC, demonstrator designs are being developed in the field of telecommunications and multimedia. The design flow, shown in Figure 9, relies on the use of C++ to capture the entire design path from system level exploration to detailed hardware design. The SLIF standard has been applied in this context to get a better view on the following IP-related issues:

1. In demonstrator-design there is a need to concentrate on the innovative bits of an application. The VC component model, as well as the interface-based design mechanism used in the SLIF standard, captures this very well.
2. The documentation methods of the SLIF standard also cope with high level behavioural models. This allows early assessment an application to an external partner.
3. The same documentation methods also allow easier demonstration of the results from the design methodology research at IMEC.

The driver application that was used in the pilot project was a digital gain control task out of a modem. This VC performs a constant multiplication of

330

an input value with a parameter. The parameter is also programmable. Despite the simplicity of the behaviour there is a large amount of ambiguities that show up at the interfaces of the task.

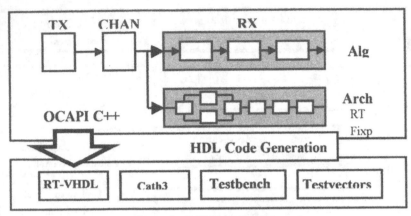

Figure 9. OCAPI methodology

The creation of a SLIF document enables the following kinds of questions to be answered: under what conditions can the parameter be updated, or when can the input be read, or what interface signals govern consumption and production of data?

The SLIF standard promotes interface-based design by using documentation *layers*. Each layer describes a VC at one level of abstraction, and each can use different execution and communication semantics. All layers fit on top of each other through refinement. This way, the initial layer expresses the most abstract view of a VC, while subsequent layers elaborate the abstract views into more concrete views.

A complete description of this example along with results is available on-line[15].

7. SUMMARY

The SLIF standard as released provides a general framework and methodology that can be used in the design process and to assist in the comprehension of interfaces, especially those exposed by Virtual Components. The usage and benefits of SLIF has been demonstrated by two very different examples. These examples clearly demonstrate that the original goals for the document have been achieved. We believe that the combination of data types, ports, attributes and transactions is an effective way to define an interface, but of more importance is the linking of these descriptions at multiple levels of abstraction, so that the design intent can always be seen.

References

[1] Virtual Socket Interface Alliance Press Release "VSIA Releases 1 1.0 Standard". In Integrated System Design, March 30, 2000

[2] "System Level Interface Behavioral Documentation Standard – Version 1.0," System Level Design, Interface Subgroup Development Working Group, Virtual Socket Interface Alliance, April 2000

[3] R. Goering, "VSIA Tackles Interface for Virtual Components – First System Level Spec Rolls". EE Times, pp. 66, April 10, 2000

[4] B. Bailey, G. deJong, P. Schaumont, C. Lennard. "Interface Based Design: Using the VSI System-Level Interface Behavioral Documentation Standard", FDL 2000, September 2000

[5] J. Rawson, A. Sangiovanni-Vincentelli., "Interface Based Design," Proc of Design Automation Conference., June 1997

[6] Virtual Socket Interface Alliance, "On-Chip Bus Attributes – Version 1.0," On-Chip Bus Development Working Group, 1998

[7] IEEE Std 1164-1993 IEEE Standard Multivalue Logic System for VHDL Model Interoperability (Std_logic_1164)

[8] Open SystemC http://www.systemc.org

[9] CynApps, Inc, http://www.cynapps.com

[10] CoDesign Automation Inc, http://www.co-design.com

[11] C Level Design Inc, http://www.cleveldesign.com

[12] Frontier Design, http://www.frontierd.com

[13] Accellera "C" Working Group, http://www.eda.org/alc-cwg

[14] A. Niemegeers; "System Level Design at Work". Medea Conference on System Level Design, September 1999

[15] IMEC VSIA Pilot Project Pages, http://www.imec.be/reuse/experience/cmult_1.2/

Virtual Component HW/SW Co-Design
From System Level Design Exploration to HW/SW Implementation

Frank Schirrmeister, Stan Krolikoski
Cadence Design Systems, Inc., USA

Key words: HW-SW co-design, SOC, system-level design, function-architecture co-design, communication refinement, design chain, communication synthesis

Abstract: As wireless, multimedia, communications, and other embedded systems converge and increase in capability, the task of system and implementation designers to efficiently design and verify embedded systems grows exponentially. Overall, the compounding complexities of chip design, silicon process, SoC system context and complete end to end verification are presenting new system design challenges. They have focused attention on tools supporting hardware-software co-design using Intellectual Property (IP) based design techniques at the system level Using methodologies and tools described in this paper, users will dramatically increase predictability and productivity of their SOC hardware/software design flows enabled by advanced reuse of system level IP. The techniques described in this paper will also enable efficient communication between system level and implementation level design teams

1. INTRODUCTION

1.1 Overview

While IP reuse from the Register Transfer Level (RTL) down to implementation often finds efficient use in today's SOC designs, the task of efficient evaluation and configuration of IP typically uses ad hoc methods. In particular, system level design evaluations attempt using models, which were intended for verification only. Typically, these prove to be too slow for evaluation and configuration.

P.J. Ashenden et al. (eds.), System-on-Chip Methodologies & Design Languages, 333–342.

In this paper we will discuss techniques, which do allow efficient system level exploration of SOC designs while avoiding the traditional simulation of implementation level models. We will describe methodologies and tools, which allow system and implementation designers to cooperate in gradually refining abstract system level descriptions by adding appropriate implementation level information. We furthermore will introduce a methodology and tools to perform communication synthesis to synthesize the HW/SW "glue" for the communication between re-usable IP blocks. Finally, we will describe methodology and tools, which automatically export the complete integrated hardware/software implementation of the system. We will also describe how to customize this flow for different targets and tools, for example different processors, real-time operating systems, debuggers, and hardware implementation tools.

1.2   The future SOC Landscape

For silicon providers the average cost of a high end ASSP and the cost of fabrication and masks have increased significantly. Looking at five to ten times return on the development cost, formerly attractive chip volumes are likely not be able to satisfy the return on investment needs in the future. Overall, the compounding complexities of chip design, silicon process, SoC system context and complete end to end verification are clearly increasing time to market for designs which are starting "from scratch".

This stands in direct contrast to market requirements in these market segments, which require short design cycles in the order of a few months. Particularly, derivative designs require quick feature upgrades without modifying the basic topology of the design.

1.3   SOC Integration Platforms

The compounding complexities of today's chip design, silicon process, SoC system context and complete end-to-end verification are presenting new challenges to system level design. Several sources ([4], [12]) suggest the notion of a SOC Integration Platform as a viable solution to those challenges. According to [4] "an integration platform for SOC design is a high productivity design environment which specifically targets a product application domain, and which is based on a VC reuse, mix and match design methodology".

In Figure 1 an example of the design content of a SOC Integration Platform is shown. The different IP blocks in hardware and software today will be authored using a variety of different description techniques. The involved processor can be designed directly in VHDL or Verilog using ad hoc techniques. Protocol and application software might be authored using the Specification Description Language (SDL) or using the Unified Modeling Language (UML, see [7], [8], [9] and [10]). Dataflow components like MPEG decoders will be defined using dedicated tools for signal processing like the Cadence signal processing work system (SPW, see also [5]). The Mathworks Matlab™ can serve to define additional IP required involving continuous time modules. In the recent past C and C++ with appropriate class library extensions have been suggested for C++ based hardware design and might result in additional Verilog and VHDL code to be implemented.

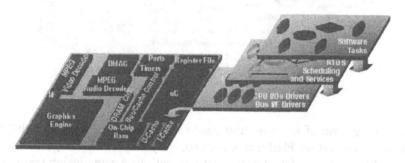

Figure 1. An SOC design example

2. DESIGN FLOWS REVISITED

As seen in the example given above today different IP sources will have to be combined to result in a integrated SOC design. Hence, two main design challenges can be derived from this example:

At first the different IP blocks have to be implemented in an optimized fashion. In this process of "IP Authoring" designers would translate embedded system requirements into an abstract system level model. Abstract, un-timed C++ or description techniques as used in SDL or Cadence SPW will ideally lead to a executable system level specification.

These system level descriptions can be used to assess and analyze the suitability and configuration of algorithms used in the system context, for example the algorithms defining the channel decoding in a cell-phone.

From this system level description the designer will then add implementation information which is required to refine the model from the abstract, un-clocked system level to the clocked implementation level. Figure 2 depicts on the left side a generalized IP Authoring design flow for hardware. For software modules Telelogic TAU and Rational UML/RT support analogous techniques to derive from a system level software description the actual software implementation, which then can be directly used in an embedded system.

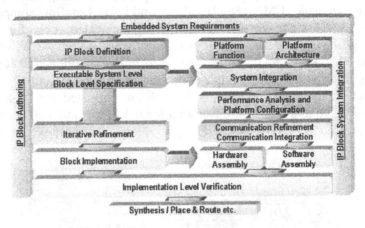

Figure 2. IP Authoring and Integration

Assuming a set of software and hardware IP libraries being in existence for a SOC Integration Platform a second, very import challenge has to be addressed for today's SOC design – IP Integration. Again, design teams will start with embedded system requirement definitions. For SOC Integration Platforms the platform functionality has to be defined, the notion of "what" the design does. As SOC Integration Platforms are typically application specific, they might enable a 2.5G or 3G wireless application, a multimedia design or a automotive application. Furthermore the platform architecture will be defined, the notion of "how" the application can be implemented using architectural components like busses, memories, processors, dedicated hardware components combined with software protocol stacks, application software and real time operating systems (RTOS).

Function and architecture together define the integrated system, on which for the integration aspects the system performance, energy consumption and cost are important decision factors. A successful co-design tool has to therefore support the assessment of performance impacts of different implementation choices for IP prior to the system implementation. This allows, based on above mentioned decision criteria, efficient system platform configuration at the system (pre-implementation) level.

In IP Integration flows design teams also do system refinement to move the design from the system to the implementation level. While the IP blocks itself ideally stay unmodified, refinement now refers to communication between IP blocks. Both functional refinement (e,g, refining how a 53 byte ATM cell has to be transmitted using six 8-byte bursts plus header transmission) and architectural refinement (e.g. refining that a hardware software communication is using a interrupt scheme with shared memory) have to be performed. Finally hardware and software have to be assembled to allow prior to SOC tape out implementation level co-verification of hardware and software. At this level of abstraction design teams will concentrate on hardware/software interface verification, as simulation speed does typically not allow complete system simulation.

Figure 2 depicts a complete design flow for both IP authoring and implementation. While in traditional design flows "system integration" mostly only happened at the implementation level, the complexities of today's and future SOC designs require "true" system level integration a higher levels of abstraction. This is indicated with the upper arrow integrating abstract system level specifications.

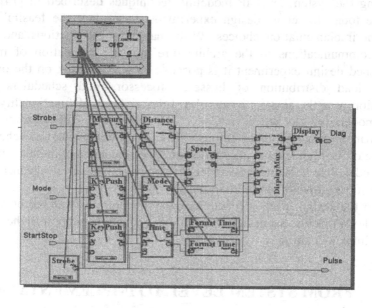

Figure 3. A Cycle-Computer design Example

3. SYSTEM LEVEL DESIGN EXPLORATION

In [2], [13], [14] efficient system level exploration methodologies for SOC designs have been proposed which allow early design trade off assessments while avoiding the traditional simulation of implementation level models. These technologies for efficient system level performance modeling via annotation of performance models to timing-free functional models, pre-implementation software estimation and efficient system exploration are available today as part of the Cadence VCC environment for virtual component co-design.

Figure 3 depicts a design example, which we will use for the discussion in this paper. It implements a cycle-computer, which is using the strobes produced by the rotation of a bicycle-wheel to compute distance, speed, acceleration etc. The functionality of this design is decomposed into hierarchical blocks of functionality implementing the different product functions. The top part of Figure 4 shows the implementation architecture on which this product functionality will be mapped. It is using a processor with a real time operating system and a dedicated hardware implementation. The communication between the architectural components is established using data and interrupt busses.

Using the system level IP modeling techniques described in [14] it is possible today to set up design experiments to explore the feasibility of different implementation choices. With mappings of functions and inter block communications to the architecture and the simulation of the so established design experiment it is possible to get feedback on the overall system load distribution of busses, processors and schedulers. The simulation uses the abstract system level descriptions of functionality (e.g. SPW models, un-clocked C or C++, SDL) combined with *characterized* performance models representing different implementation choices.. Because of the high level of abstraction and the avoidance of implementation level models like HDL and assembler the simulation speed is very fast and a variety of experiments can be analyzed.

After the design space of function and architecture is explored using fast system level models, the design needs to be prepared for export to the actual system implementation.

4.   FROM SYSTEM LEVEL TO IMPLEMENTATION

Several techniques available today do attempt the synthesis of large block content itself. The methodology described here concentrates on exporting the integrated design of available implementation models coming with the system level models. To address the special integration aspects, communication synthesis and refinement is supported. For more details please refer to [15]

4.1   Communication Refinement

In the un-timed system level description, the user can define data types representing the level of abstraction both at which the system designer thinks and at which the specification takes place (e.g. ATM cells, GSM frames etc.). For instance, suppose a communication wire passes a 64-bit token between functions in the design. Assuming this signal has to be mapped to an architectural bus that accepts 16-bit tokens, *Communication Refinement* helps resolve this type mismatch. In this case, the designer can simply add the appropriate communication protocol that refines the 64-bit token into the appropriate 16-bit segments to be transmitted.

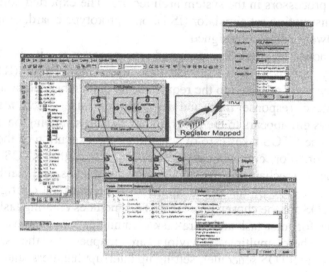

Figure 4. HW/SW communication

4.2 Communication Synthesis

The VCC environment synthesizes the communication "glue" that connects the implementation version of hardware and software "virtual components", pieces of intellectual property (IP). Most of what VCC exports is function calls and component instantiations from the IP database that implement what was modeled in the VCC behavioral blocks. VCC is responsible for generating or instantiating the code and circuitry that implement the VCC communications semantics as appropriate for each context.

Figure 4 shows an example of a communication from hardware to software. Behaviors mapped to hardware, that need to communicate with behaviors mapped to software, typically do so either by raising interrupts or by setting flags that the software periodically polls. VCC export allows the user to select which scheme to use.

When an interrupt is raised by the hardware, an interrupt service routine (ISR) will be called. The VCC environment supports a large variety of 17 communication patterns between HW/HW, SW/SW, SW/HW and HW/SW.

4.3 Software Export

The software export feature produces the parts of the design that are mapped to processors in the system architecture. The exported software can run on an instruction set simulator (ISS), on a prototype board, or ultimately on the hardware being co-designed.

With more functionality assigned to processors, it is becoming increasingly common to use a real-time operating system (RTOS), an operating system optimized to the requirements that embedded systems have for guaranteed response time and low memory use. Each *task* in the operating system appears to the programmer as if it is running on a separate processor. The RTOS takes care of scheduling the execution of the tasks on the processor. For communication between tasks, the RTOS provides communication mechanisms such as semaphores, queues, and mailboxes.

Software export of the user's application consists of configuring the selected RTOS for the chosen processor. This includes creating tasks, adding mechanisms for inter-task and intra-task communication, synthesizing static schedulers where multiple behaviors are mapped to the same task, configuring the RTOS scheduler, setting up interrupt handlers, and setting up counters and timers.

4.4 Hardware Export

Hardware export generates a pin-accurate, structural HDL description of the architecture. It also exports register-transfer level (RTL) HDL descriptions for the behaviors mapped to custom blocks in the architecture. These descriptions can then be fed into downstream tools such as chip floor planning, simulation, and synthesis.

The 'relaxed' high-level VCC architecture is refined by hardware export to the pin-accurate level. The refined architectural instances are connected according to IP-dependent expansion rules. The architecture is then exported as a pin-accurate structural Verilog or VHDL description.

A high-level specification of the architecture is all that is required for system integration and evaluation. However, for design export purposes this abstract view of the architecture needs to be refined in order for a structural description to be exported for use by the downstream tools. Each of the high-level architectural instances, which have a small number of pins representing complex bus connections, is replaced by a corresponding expanded, pin-accurate architectural instance—this information is obtained from the IP library. Architecture expansion is also applied to buses. High-level representations of the buses are replaced with individual bus nets and bus support logic, including decoders, arbiters, and interrupt controllers. Expansion rules define how the refined architecture instances and buses connect to each other.

4.5   Test-bench Export

VCC allows the user to verify the correspondence between the VCC design and selected parts of the exported implementation design. This partitioning of the design speeds up simulation by simulating only certain parts of the design at the detailed implementation level (see Figure 8).

Implementation hardware probes are HDL blocks that are connected up in the hardware part of the implementation so that they can read from the points corresponding to probes in the VCC model, and write into the same results database to which high-level VCC simulation results are written. Sources are HDL blocks that can be used to read previously captured probe data from the results database and inject it into the hardware simulation.

VCC can cause the implementation software debugger or the HDL simulator to generate tracing information that corresponds to the high-level probes that the user inserted in the original VCC design. A probe is a type of test bench block for observing the system, which is handled specially by the VCC simulation environment, and simulates in the implementation with little perturbation of the performance of the implementation.

5.   CONCLUSION

We have briefly described a methodology and tools, which allow the design-export from system level descriptions to the implementation level. The next tool chain used from here on would be Co-Verification. Figure 9 shows an example of actual data probed in a VCC simulation followed by the equivalent data probed in a co-verification tool for the design exported by VCC. An important aspect in this comparison is that 40 sec test-bench simulation took only several minutes in VCC, while the same simulation took 11 hours in Co-Verification. Using the abstract performance simulation techniques, a large variety of design alternatives can be explored.

The technologies mentioned in this paper have been developed as part of the Felix Initiative, which did build upon the joint research work conducted by the University of California at Berkeley, Cadence Berkeley and European Labs, and Magneti Marelli. Version 1.2 of the Virtual Component Co-

342

Design tools is available since January 2000. Partners and Early Adopters for this development include Advanced RISC Machines, Ltd., BMW, debis Systemhaus, Magneti Marelli S.p.A., Motorola SPS; National Semiconductor Corporation, ST Microelectronics and Telefonaktiebolaget LM Ericsson.

References
1 G. Martin and Sanjay Chakravarty, "A New Embedded System Design Flow based on IP Integration", DATE99, User Forum
2 G. Martin and B. Salefski, "Methodology and Technology for Design of Communications and Multimedia Products via System-Level IP Integration", DATE98
3 S. Chakravarty, "IP Modeling for a GSM Handset in the VCC Design Environment", DATE99, User Forum
4 G. Martin, H. Chang, L. Cooke, M. Hunt, A. McNeilly, L. Todd, "Surviving the SoC Revolution: a Guide to Platform-Based Design", Kluwer Academic Press
5 Sanjay Chakravarty, Stan Krolikoski, Grant Martin, "DSP Software Estimation using characterized kernel functions", DSP Deutschland, September 1999.
6 F. Balarin, M. Chiodo, P. Giusto, H. Hsieh. A. Jurecska, L. Lavagno, C. Passerone, A. Sangiovanni Vincentelli, E. Sentovich, K. Suzuki and B. Tabbara, "Hardware-Software Co-Design of Embedded Systems", Kluwer Academic Publishers, Dordrecht, The Netherlands, 1997
7 G. Booch, J. Rumbaugh, I. Jacobson, The Unified Modeling Language User Guide, Addison-Wesley, 1999
8 G. Booch, J. Rumbaugh, I. Jacobson, The Unified Modeling Language Reference Manual, Addison-Wesley, December, 1998
9 Bran Selic and Jim Rumbaugh, "Using UML for Modeling Complex Real-Time Systems", white paper, ObjecTime, March 11, 1998
10 B. Selic, "Turning clockwise: using UML in the real-time domain," Communications of the ACM, vol. 42, number 10, pp. 46-54, October, 1999.
11 Open Model Interface Standard, see http:// www.cfi.org/OMF/.
12 Alberto Sangiovanni-Vincentelli and Alberto Ferrari, "System Design – Traditional Concepts and New Paradigms", Keynote at ICCD99
13 Frank Schirrmeister, Stanley Krolikoski, "Virtual Component Co-Design – Facilitating a Win-Win Relationship between IP Integrators and IP Providers"; IP Conference, November 1999, Edinburgh
14 Frank Schirrmeister, Stanley Krolikoski, "Modeling Techniques for Evaluation and Configuration of System Level Intellectual Property"; International Workshop on IP Based Synthesis, December 1999, Grenoble
15 Mark Baker and Eamonn O'Brien-Strain, "Co-Design Made Real: Generating and Verifying Complete System Hardware and Software Implementations", Embedded Systems Conference, San Jose, CA; September 1999; Paper #520